ROUTLEDGE LIBRARY EDITIONS: URBANIZATION

Volume 4

INDUSTRIALIZATION AND URBANIZATION IN LATIN AMERICA

INDUSTRIALIZATION AND URBANIZATION IN LATIN AMERICA

ROBERT N. GWYNNE

Routledge
Taylor & Francis Group
LONDON AND NEW YORK

First published in 1985 by Croom Helm Ltd

This edition first published in 2018
by Routledge
2 Park Square, Milton Park, Abingdon, Oxon OX14 4RN

and by Routledge
605 Third Avenue, New York, NY 10017

Routledge is an imprint of the Taylor & Francis Group, an informa business

British Library Cataloguing in Publication Data
A catalogue record for this book is available from the British Library

ISBN: 978-0-8153-8014-6 (Set)
ISBN: 978-1-351-21390-5 (Set) (ebk)
ISBN: 978-0-8153-7902-7 (Volume 4) (hbk)
ISBN: 978-1-351-21698-2 (Volume 4) (ebk)

Industrialization and Urbanization in Latin America

ROBERT N. GWYNNE

CROOM HELM
London & Sydney

Croom Helm Ltd, Provident House, Burrell Row,
Beckenham, Kent BR3 1AT
Croom Helm Australia Pty Ltd, Suite 4, 6th Floor,
64–76 Kippax Street, Surrey Hills, NSW 2010, Australia

British Library Cataloguing in Publication Data

Gwynne, Robert N.
Industrialisation and urbanisation in Latin America.
1. Latin America — Industries
I. Title
338.098 HC123

ISBN 0-7099-1110-6

Photoset in English Times by Pat and Anne Murphy,
10 Bracken Way, Walkford, Christchurch, Dorset

Printed and bound in Great Britain
by Billing & Sons Limited, Worcester.

CONTENTS

List of Figures
List of Tables
Preface

Part One: The Process of Industrialisation

1. Latin America in the World Economy 3
2. Industrial Policies 22
3. Industrial Structure and Technological Change 35
4. The Institutional Framework 49

Part Two: Spatial Perspectives on Industrialisation

5. Regional Economic Integration 71
6. The Spatial Centralisation of Industry 82
7. Processes of Industrial Decentralisation 103

Part Three: Industrialisation and Urbanisation

8. Historical Perspectives on Urbanisation 135
9. Industrialisation and Urbanisation 146

Part Four: Industrialisation and Regional Development

10. Theoretical Perspectives on Industrialisation and Regional Development 171
11. Industrialisation and Regional Development in Chile 194
12. Conclusion: The Importance of Industrial Growth 237

Bibliography 243
Index 254

FIGURES

1.1 Composite Commodity Price Index, 1948–1982 6
1.2 Latin American Per Capita Debt, 1983 15
2.1 Notional Relationship Between Output of a Vehicle Assembly Operation and Per Unit Costs in Latin America 27
2.2 Brazilian Exports, 1971–1981 29
2.3 Stages in the Product Life Cycle 31
6.1 Provincial Variation in Annual Average Salaries for White Collar Workers in Chile, 1967 90
6.2 Provincial Variation in Annual Average Wage Rates for Blue Collar Workers in Chile, 1967 91
6.3 Provincial Variation in Annual Average Salaries for White Collar Workers in Chile, 1978 92
6.4 Provincial Variation in Annual Average Wage Rates for Blue Collar Workers in Chile, 1978 93
7.1 The Location of the Motor Vehicle Industry in Chile 108
7.2 The Mexican Border Industrialisation Programme 122
7.3 The Venezuelan Axial Belt 125
7.4 Spatial Divisions of Venezuela's Industrial Decentralisation Programme 126
7.5 The Industrial Core of Brazil 128
8.1 Chilean Railways, 1900 141
9.1 Chile's Central Region 151
10.1 Williamson's Model of the Relationship Between Economic Development and Regional Disparities Through Time 173
10.2 The Regional Development Strategy of Chile's Christian Democrat Government, 1964–1970 183
11.1 Chile: Employment Change by Industrial Sector, 1928–1957 200
11.2 Chile: The Manufacturing Region of Concepción in 1957 205

TABLES

1.1 Changes in the structure of Latin American exports, 1960–80 4
1.2 Changes in the structure of Latin American imports, 1960–80 5
1.3 Structure and level of net inflow of external resources to Latin America, 1961–78 11
1.4 Latin American Debt, 1983 13
1.5 World distribution of manufacturing output and trade by region, 1955 and 1975 17
2.1 Distribution of motor vehicle production by country and firm in Latin America, 1977 26
3.1 Growth rates of population and the manufacturing product in Latin America, 1950–78 36
3.2 Manufacturing and other indices for Latin American countries, 1950 37
3.3 Manufacturing and other indices for Latin American countries, 1978 38
3.4 Industrial structure and size of country in Latin America, 1950–75 40
3.5 Sectoral growth and size of country in Latin America, 1950–75 41
4.1 Company interests of Chile's six leading conglomerates, 1978 53
4.2 Percentages of assets of the largest 300 manufacturing firms in Brazil and Mexico held by US, other foreign, and national companies, 1972 54
4.3 Changes in the investment patterns of multinational enterprise in Brazil, Mexico and other Latin American countries 57
4.4 Distribution of foreign direct investment in Brazilian manufacturing by sector, 1979 58
5.1 Allocations, assembly and co-production agreements within the Andean Pact Automobile Programme, 1983 78
6.1 Primacy in Latin America 86
6.2 Size of city and income levels in Mexico 86

6.3	The relationship between city size and retail prices, Chile, 1970	87
6.4	The changing rank of Chilean provinces between 1967 and 1978 in terms of salaries and wages	89
7.1	Location of plants producing components for the Chilean motor vehicle industry, 1969	109
7.2	Location of the important vehicle component plants of Chile, 1971, according to industrial type	110
7.3	Location of suppliers of Fiat's plant in Belo Horizonte, Minas Gerais	112
8.1	Two indices of the percentage of the total population of each world region that is urban	135
8.2	Primacy in Chile, 1813–1920	143
9.1	Ratio of manufacturing to agricultural gross domestic product, 1979	147
9.2	Manufacturing workers as a percentage of the total non-agricultural workforce for urban areas in Chile, 1970	150
9.3	Characteristics of the two circuits of the urban economy in Latin America	160
10.1	São Paulo state: comparison of average annual population growth rates	180
10.2	Latin American companies with 1982 sales greater than one billion dollars	188
10.3	Urban population as a percentage of total population, 1981	191
11.1	Chile: immigrants as industrial entrepreneurs (owners, partners, lessees) and their distribution by province, 1928	196
11.2	Chile: the changing distribution of manufacturing employment, 1928–57	203
11.3	The centralisation of Chilean industry in the Santiago metropolitan area by sector, 1957	204
11.4	The progressive integration levels imposed on motor vehicle companies producing in Chile, 1963–9	211
11.5	Chile: increases in the proportion of foreign investment in certain manufacturing sectors, 1967–9	212
11.6	Foreign investment in the hundred principal private companies of Chile in 1969	213

11.7 The changing distribution of manufacturing employment, 1957–67 217
11.8 The spatial centralisation of selected industrial sectors in Santiago province in 1967 based on employment and value-added totals 218
11.9 The changing distribution of manufacturing employment in medium and large plants (over 50 employees) between 1967 and 1973 219
11.10 Changes in industrial production and employment in Chile by sector, 1973–80 224
11.11 Chilean industrial sectors classified according to change in levels of production and employment between 1973 and 1980 226
11.12 Chile: imports of manufactured consumer goods, 1973–83 229
11.13 Chile: indices of manufacturing growth, 1979-June 1984 230
11.14 Chile: changes in regional manufacturing employment, 1973–80 232
11.15 Chile: changes in regional manufacturing production, 1973–80 233

TO MY MOTHER AND FATHER

PREFACE

In the 1950s, industrialisation was seen as a panacea for the development of less developed countries in general and Latin American countries in particular. Geographical perspectives on industrialisation were provided by Alan Mountjoy's book, *Industrialisation and Underdeveloped countries*. In retrospect, too much was claimed for the strategy of industrialisation. Investment in industrial expansion alone could not provide for rapid growth and development. Multipliers from industrial investments did filter through the urban economies of less developed countries, but agriculture stagnated unless policies were also able to promote change in the rural sectors..

As with all panaceas that do not live up to their credentials, the policy of industrialisation for less developed countries was later criticised, even vilified. Lipton went so far as to blame the policy and process of urban-based industrialisation as the root cause of why people stay poor in the rural areas of less developed countries. Many academics criticised the important role it gave to multinational companies in the process of development. Others pointed to the dependence on foreign technology and products that industrial expansion brought. Many argued that investment in agriculture should come before investment in industry in the poorer countries of the world. As industrialisation did not bring the substantial increases in employment earlier expected of it, some argued that more labour-intensive forms of economic development were necessary.

Thus, the 1960s and 1970s saw the consensus of development theorists turn away from industrialisation in favour of other policies. But in the wake of the 1979–84 recession, that has affected less developed countries so much more seriously than developed countries, the benefits of industrial policies for less developed countries have once again become evident. As prices for primary commodities, still the basic export of most less developed countries, have tumbled in real terms between 1979 and 1984, the only less developed countries that have been able to maintain and improve their trading position in the world are those that have industrialised in the past three decades. These so-called newly-industrialising countries, such as Taiwan, Korea, Hong Kong, Brazil and Mexico, have been the only group of less developed countries to expand the value of their exports between 1980 and 1984; in particular, the increased export of manufactured products has been directed towards the expanding US economy between 1983 and 1984.

This book focuses on the process of industrialisation in but one of the continents of the Third World, Latin America. The aim of the book, however, is to link the distinctive process of industrialisation to wider issues of urban and regional development in Latin America. As a result, the book is divided into four parts.

Part One specifically examines the process of industrialisation in Latin America. In the first chapter, the external constraints on industrialising in a world economy of increasing protection and high interest rates are examined. Nevertheless, the advantages that industrialising countries have over non-industrialising countries are emphasised. In Chapter Two, the basic strategies of industrial development are discussed. The orientation of industry towards the domestic market (import substitution industrialisation) is set against a more outward-looking approach in which the need for generating manufactured exports is recognised. However, the historical importance of import substitution policies has operated to differentiate between large and small countries in terms of industrial growth, structure and technology. The larger countries have developed more mature industrial structures than smaller countries and have become important centres of technological expertise and adaptation. Chapter Three focuses on spatial variations in industrial structure and technology between Latin American countries. The following chapter examines the organisations that have been responsible for the process of industrialisation in Latin America, namely state firms, private national enterprises and multinational corporations — known collectively as the triple alliance.

Part Two focuses on the spatial ramifications of Latin American industrialisation. The existence of large numbers of small countries (both in terms of population and area) created the need for policies of regional economic integration in order to further industrial growth geared to an expanded regional market. Chapter Five examines the problems and principal features of such attempts at spatial integration on a regional scale. Within most countries, however, the distinctive spatial character of industrial development has been its exaggerated concentration in one metropolitan area — normally but not necessarily the capital city. Chapter Six pursues a locational analysis of this spatial phenomenon. Governments have attempted to reduce this 'natural' phenomenon of spatial centralisation by various means and Chapter Seven examines the success of these schemes of industrial decentralisation.

In the growth of nineteenth century Europe, there was an intimate connection between the growth of towns and the process of industrialisation. Owing to the very different type of industrial growth in the latter half of the twentieth century, comparable connections are not evident in Latin America. However, industrialisation is still the single most powerful economic force behind the tremendous expansion of Latin American cities in recent years. Chapter Nine explains these links between modern industrialisation and urbanisation in Latin America after Chapter Eight has put the Latin American city into historical perspective. The connections between industrialisation, the urban hierarchy and the large city are examined as are various theories that have sought to conceptualise about the relationship between 'capital-intensive' industry and 'labour-abundant' cities.

It has been argued that industrial growth and its geographical distribution is a principal cause of increasing disparities in income between regions within Latin American countries. Chapter Ten examines the theoretical and empirical bases of the argument. Chapter Eleven concludes the book by providing a detailed case study of the complex relationships between industrial and regional development in Chile.

Material for the book has been acquired from five separate research investigations and trips to Latin America — in 1973/74, 1977, 1979, 1980 and 1984. For helping to finance these investigations, I would like to thank the Parry Commission of the Department of Education and Science, the University of Birmingham, the British Council and the Economic and Social Research Council. In my main study area of Chile, I would like to express thanks to the following institutions for their assistance: the Departments of Geography of the University of Chile, the Institute of Urban Studies and the Department of Geography of the Catholic University of Chile, and the Department of Geography of the University of Santiago. I am grateful for the help and good will of a great many people in Latin America but in particular Anibal, Nono, Ivan, Hernan, Maria Eliana, Fernando, Hugo and Alfredo.

Most of all I would like to thank my wife, Maruja, for being an excellent Spanish teacher in the early days and for her generous, but nevertheless acute, perceptions of a Latin American reality far removed from the ideologies that seem to prosper in and about the continent.

I am grateful for the assistance of colleagues in British Universities who have commented upon parts of the book or helped in research problems — namely Harold Blakemore, Sue Cunningham, John Dickenson, Alan Gilbert, George Philip, David Preston, and Clifford Smith. I would also like to thank all colleagues at the Department of Geography in the University of Birmingham. In particular I am indebted to the design and drawing abilities of Tim Grogan and the expert typing of Lynn Ford and Judy Astle.

PART ONE

THE PROCESS OF INDUSTRIALISATION

1 LATIN AMERICA IN THE WORLD ECONOMY

With the exception of Cuba, the economies of Latin America are closely interlinked into the world capitalist system. With this one exception, the countries of Latin America have perceived their advantage in remaining and participating in the Western economic system. The relative stagnation of the Cuban economy since 1958 and its heavy dependence on the Soviet government have not stimulated further resignations from the Western capitalist system. Indeed, Nicaragua maintains close trading and technological links with West European countries. Generally speaking Latin American countries realise that they receive net benefits from interaction and trade with the industrialised countries of the West and those less developed countries that participate in the Western trading system.

As a result, the operation of the world capitalist system significantly affects both the pace and rhythm of economic growth in each Latin American country. It is the purpose of this chapter to assess some of the mechanisms of the world capitalist system that affect Latin American economies and in particular the process of industrialisation. There are two principal external mechanisms that presently have a fundamental significance for the operation of Latin American economies — international trade and capital flows.

International Trade

Since the early nineteenth century and the independence of most Latin American countries from Spanish rule, trade between Latin America and the industrialised countries has maintained a very distinctive pattern. On the one hand, there has been the export of fuels, minerals, metals, agricultural and other primary products from Latin America. On the other hand, there has been the import of manufactured products into Latin America. Tables 1.1 and 1.2 demonstrate that such a pattern was still strongly in evidence in 1960. With the exception of Uruguay and Brazil, primary commodities still accounted for over 88% of the exports of all Latin

Table 1.1: Changes in the Structure of Latin American Exports, 1960–80 (Percentage share of merchandise exports)

	Fuels, minerals and metals		Other primary commodities		Manufactures	
	1960	1980	1960	1980	1960	1980
Oil-exporting countries						
Venezuela	74	98	26	–	–	2
Mexico	24	39	64	22	12	39
Ecuador	–	56	99	41	1	3
Lower middle-income countries						
Bolivia	–	86	–	11	–	3
Honduras	5	7	93	81	2	12
El Salvador	–	5	94	59	6	36
Nicaragua	3	3	95	83	2	14
Guatemala	2	6	95	70	3	24
Peru	49	64	50	20	1	16
Colombia	19	3	79	77	2	20
Costa Rica	–	1	95	65	5	34
Paraguay	–	–	100	88	–	12
Upper middle-income countries						
Panama	–	24	–	67	–	9
Brazil	8	11	89	50	3	39
Argentina	1	6	95	71	4	23
Chile	92	59	4	21	4	20
Uruguay	–	1	71	61	29	38

Source: World Bank, *World Development Report 1983* (Oxford University Press, New York, 1983)

American countries and manufactures for over 70% of their imports.

Several factors have changed this trading pattern in the last twenty years. The best known is the rise in the price of oil, which has meant the doubling in the importance of this commodity in international commerce, from around 10 per cent of total world exports in 1970[1] to more than 20 per cent in 1980.[2] Three major oil-exporting nations exist within Latin America — Venezuela, Ecuador and Mexico, of which the latter two have increased their trading importance within Latin America in the last two decades.

A second important change in world trade has been the fall in the share of less developed countries in world exports of primary products other than oil. Less developed countries accounted for

Table 1.2: Changes in the Structure of Latin American Imports, 1960–80 (Percentage share of merchandise imports)

	Fuels		Food and other primary commodities		Manufactures and Machinery	
	1960	1980	1960	1980	1960	1980
Oil-exporting countries						
Venezuela	1	2	28	20	71	78
Mexico	2	2	14	15	84	83
Ecuador	3	1	22	12	75	87
Lower middle-income countries						
Bolivia	—	1	—	12	—	87
Honduras	9	16	16	12	75	72
El Salvador	6	18	23	22	71	60
Nicaragua	10	20	14	16	76	64
Guatemala	10	24	19	15	71	61
Peru	5	2	21	25	74	73
Colombia	3	12	23	18	74	70
Costa Rica	6	15	19	13	75	72
Paraguay	—	24	—	14	—	62
Upper middle-income countries						
Panama	10	31	16	11	74	58
Brazil	19	43	27	16	54	41
Argentina	13	16	14	13	73	77
Chile	—	21	—	18	—	61
Uruguay	24	29	51	15	25	56

Source: World Bank, *World Development Report 1983* (Oxford University Press, New York, 1983)

less than 36 per cent of exports of raw materials in 1978, compared with a share of almost 44 per cent in 1960.[3] This phenomenon has been the principal cause of the decline in Latin America's share of total world trade since 1960.

Thirdly, prices for non-fuel commodities have been both declining and unstable in the last thirty years. Figure 1.1 shows that non-fuel commodities as a whole have declined in value by over a third from the early 1950s to the early 1980s. At the same time, considerable price instability exists alongside the general decline, as price behaviour in the early 1970s demonstrates (see Figure 1.1). In certain commodities, the decline (and instability) of prices has been greater than this composite average. In 1982, the real value of copper was only 20% of its 1970 value,[4] a fact that has caused

Figure 1.1: Composite Commodity Price Index, 1948–82

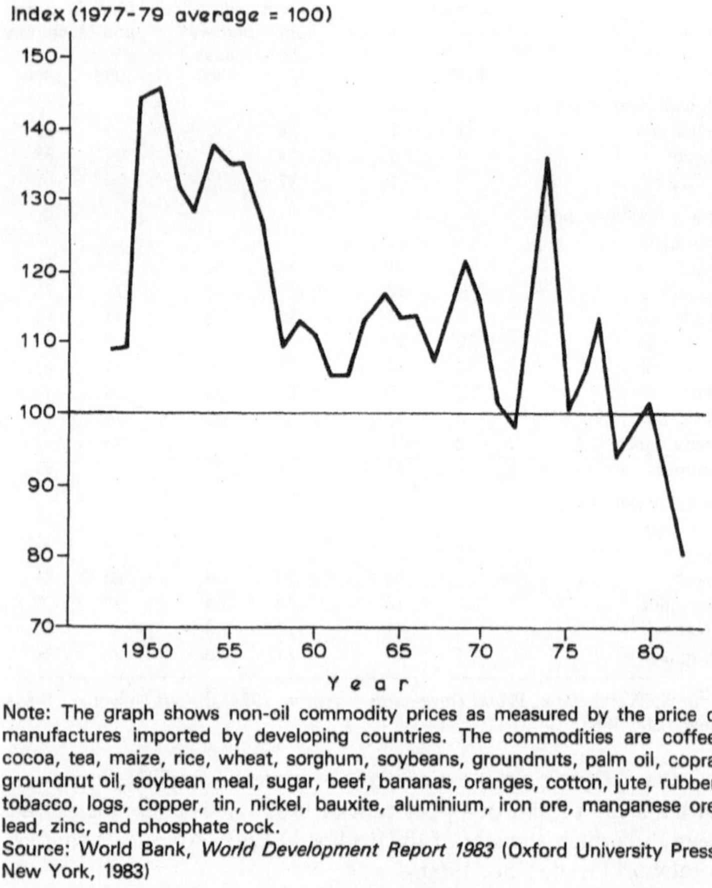

Note: The graph shows non-oil commodity prices as measured by the price of manufactures imported by developing countries. The commodities are coffee, cocoa, tea, maize, rice, wheat, sorghum, soybeans, groundnuts, palm oil, copra, groundnut oil, soybean meal, sugar, beef, bananas, oranges, cotton, jute, rubber, tobacco, logs, copper, tin, nickel, bauxite, aluminium, iron ore, manganese ore, lead, zinc, and phosphate rock.
Source: World Bank, *World Development Report 1983* (Oxford University Press, New York, 1983)

great stress for the world's copper exporters.

With declining prices and world market share in non-fuel commodities, it has been necessary for those Latin American countries without large oil surpluses to increase their exports of manufactures, the most dynamic sector of world trade. Indeed, the share of manufactures within total exports from the region has increased from 3 per cent in 1960 to over 18 per cent in 1978. By 1980, the manufacturing share was over 30 per cent of total exports in five countries — Brazil, Mexico, Uruguay, El Salvador and Costa Rica (see Table 1.1).

According to international trade theory, Latin American countries should concentrate on the production and export of labour-intensive goods because they are 'labour abundant' nations and labour costs are relatively low in international terms. At the same time, Latin American countries should import capital-intensive goods from the 'capital-abundant' industrialised nations.[5] There are, however, several factors that make the future growth of manufactured exports from Latin America more difficult than such idealistic theory would predict. Some of these relate to mechanisms external to Latin American economies.

One major problem is that of protectionism in the developed countries. In the 1960s and 1970s the developed countries became more open to imports of manufactured goods from less developed countries. For example, between 1971 and 1976, the EEC, Japan and the United States implemented the General System of Preferences (GSP) for imports from developing countries. These have certainly been important in facilitating the export of manufactured goods from less developed countries. For example, the EEC's Generalised System of Preferences covers between 3 per cent and 4 per cent of the total imports from Latin America but as much as 30 per cent of its manufactured goods imported from that region.[6] Nevertheless, it has been argued that the impact of the GSP has been limited. Instead of complete, permanent and generalised tariff reductions, the GSP allows for partial, temporary and discretionary reductions confined to non-sensitive products. The GSP also allows for import quotas which have often proved very restrictive. Certainly the 'partial, temporary and discretionary' nature of the GSP make it easy for the developed countries to reduce whatever benefits the GSP has provided for the developing countries. Furthermore, there is evidence of distinctly more protectionist stances by the developed countries in the 1980s. The non-signing of the Multifibre Agreement by the EEC in 1981 is one example of this.

It has been argued that Latin American countries can be doubly penalised by a more protectionist stance from the developed countries as they have never developed a special relationship with any of the groupings of the developed world. For example, they have no special relationship with the EEC whose markets for manufactured goods are open, without tariff or quota restrictions, to exports from the 52 developing countries of Africa, the Caribbean and the

Pacific Islands who were signatories of the Lomé Convention. Japan's system of regional preference favours the Far Eastern Countries while the United States has not promoted such a system — at least not until the 1981 special arrangements for the Caribbean area. In this way, Latin American manufactured exports can be prevented from entering the markets of developed countries. One example of this was the decision of the West German government under pressure from its car unions to deny entry to finished Brazilian Volkswagen vehicles. Brazilian-produced Volkswagens were in some cases cheaper than domestically-produced vehicles but their entry into West Germany and the EEC was denied.

For this reason, the majority of Latin America's manufactured exports go to other less developed countries. In 1975, for instance, 90 per cent of all Latin American manufactured exports stayed in the region. Subsequently, Brazil's major thrust in its policy of export promotion of manufactured goods has been directed to other Third World countries. Volkswagen cars have been exported in large numbers to such countries as Nigeria, the Philippines and Chile. However, such a direction had adverse repercussions in the 1979–83 recession, in which Third World countries suffered more than developed countries. As a result, they severely restricted their manufactured imports which in turn affected Brazil's manufactured exports. This was the major reason for Brazil's decline in manufactured exports between 1981 and 1982. The more volatile nature of Third World markets for manufactured imports has become another constraint on the export promotion policy of Brazil and other Latin American countries.

A fourth potential problem is that a significant proportion of the exports of manufactured products from Latin America are carried out by multinational companies. Early 1970s data demonstrate that 30 per cent of Mexican, Colombian and Argentinian manufactured exports were made through multinationals and as much as 43 per cent of Brazil's (1969 data).[7] Multinationals export from Latin American countries when there are cost advantages for them. These cost advantages most commonly arise from either labour cost differentials (particularly for South-North trade) or from a need to increase capacity utilisation (often the incentive for increased South-South trade). Due in particular to the labour cost factor, multinationals favour the location of assembly-type industries in Latin American countries — industries, that is, that require large amounts of labour. However, as assembly-type operations become more and more capital intensive, as with the increasing use of robots in motor vehicle assembly, what will be the advantage of locating in cheap labour locations? As the development of labour-saving technology proceeds apace in the North and undercuts the unit cost production of cheap labour locations in the South, there will be increasingly more powerful arguments for multinationals to bring 'home' the production from overseas — or at least to limit severely the imports from the countries of the South, including Latin America.

International Capital Flows

Economic growth and industrialisation in Latin America are increasingly affected by the flow of international capital and its relative abundance or scarcity. In the last forty years, Latin America has experienced at least three contrasting periods of capital inflow in terms of both type and quantity.

After the Second World War, Latin America appealed for a Marshall Plan equivalent to that for Western Europe. George Marshall replied that only private funds would be available. The main reason for such parsimony was that Latin America played a very secondary role in the Cold War at that time. As a result, from 1945 to 1960, capital flows to Latin America were mainly private investments by American corporations. The flow of capital was small and closely linked to profitable corporate investments.

Towards the end of the 1950s, the American attitude towards official aid began to change. The turning point was the triumph of the Cuban revolution in 1959 and the subsequent launching of a massive aid programme under the Alliance for Progress. The Kennedy government believed that economic development and social reform, financed by US aid, would curtail the spread of revolution from Cuba. The total level of gross US economic assistance to Latin America during the 1960s was over ten billion dollars.[8] However, the level of official flows began to slow down in the late 1960s as the effects of the Cuban revolution were seen to have been contained.

By the late 1960s, many Latin American countries that had encouraged industrialisation during the 1950s and 1960s found that their access to financial resources fell short of requirements. The net inflow of external credit between 1966 and 1970 averaged only

$2.6 billion a year — one third in the form of direct corporate investment, one quarter in US aid, and one sixth from official international agencies (such as the IMF). International banks accounted for less than 10 per cent of the total.

In the 1970s, the nature of capital inflow into Latin America dramatically changed. First, the size of capital inflow substantially increased — from an average of only $2.6 billion a year between 1966 and 1970 to $21.8 billion in 1978 — an approximate tenfold increase in ten years. Secondly, the origin of the funds changed; in relative terms, the influence of direct investment and US bilateral aid substantially declined. Multinational banks came to account for the majority of foreign credit — 57 per cent in 1978 (see Table 1.3).

How and why have the multinational banks achieved such dominance in the provision of external credit to Latin America? Griffith Jones points out that since the early 1960s, most of the world's major banks have emulated other large corporations by 'going transnational' and opening up branches outside their national borders.[9] In 1960, only eight US banks had foreign branches; by 1975, 125 US banks had foreign branches. Total assets of US overseas branches jumped from $3.5 bn in 1960 to $181 bn by June 1976. Banks from Europe, Japan and the Middle East began their expansion later, but their growth in overseas lending has also been rapid.

The big surge in bank lending to Latin American can be traced to the oil price rises of 1973 and the consequent surpluses of capital accruing to some OPEC countries, such as Saudi Arabia and Kuwait. These surpluses began to be recycled through the 'Eurodollar' currency market. The amount made available through the 'Eurodollar' currency market between 1973 and 1974 has been calculated at over $60 billion. Official agencies such as the International Monetary Fund (IMF) had their funds boosted by some $19 billion during 1974–5. But the lion's share of the surplus found its way into the private commercial banking sector. The banks involved extended 60 per cent or more of their Eurocurrency credits to less developed countries and particularly to Latin American countries.[10]

This rapid increase in bank lending to Latin American countries was the result of a variety of factors. First, international banks found that the credit demand of traditional clients had slowed down as a result of the recession, at the same time that deposits from oil-exporters were growing very fast. This prompted the

Table 1.3: Structure and Level of Net Inflow of External Resources to Latin America 1961–78

	Annual Averages					
% Structure	1961–65	1966–70	1971–75	1976	1977	1978
I Net public inflow	60.2	40.1	25.2	19.6	12.0	7.3
A. Multilateral	19.5	15.7	13.4	14.4	7.4	3.1
1. Development	16.6	17.1	11.6	6.6	8.4	7.2
2. Compensatory	2.9	−1.4	1.8	7.8	−1.0	−4.1
B. Bilateral	40.7	24.4	11.8	5.2	4.6	4.2
1. US	36.9	23.6	6.8	2.6	1.7	0.8
2. Other Countries	3.8	0.8	5.0	2.6	2.9	3.4
II Net private inflow	39.8	59.9	74.8	80.4	88.0	92.7
A. Banks	2.1	9.3	43.8	61.0	48.3	56.6
B. Suppliers	7.7	13.8	2.3	3.7	5.8	9.8
C. Bonds	5.0	2.5	2.5	3.3	14.8	10.3
D. Direct Investment	25.2	33.3	26.2	12.4	20.1	16.0
III Total %	100.0	100.0	100.0	100.0	100.0	100.0
Total Actual Level (US$m)	1,575.8	2,641.3	7,561.9	15,301.5	15,637.0	21,807.2

Source: S. Griffith Jones, Transitional Finance and Latin American National Development, *IDS Discussion Paper 175*, 1982, Sussex

banks to lend to borrowers previously considered marginal. Meanwhile the public and private sectors in Latin America were keen to borrow. As already mentioned, in the early 1970s, there was a relative stagnation in the net flow of official development assistance. Furthermore, Latin American governments often preferred private bank to IMF loans as conditions were much easier. Loans could be made effective more quickly, employed in a greater variety of ways, and given without reference to the management of the national economy. Many governments preferred foreign private loans to direct investment, because the former were perceived as having fewer ties and this allowed greater autonomy for national development. Demand for foreign loans increased in particular among oil-deficit countries, where foreign funds were used to balance trading and current account deficits.

An international system of capital flows developed that seemed to benefit all parties. The OPEC countries approved of the system. Their huge surplus funds were being safely invested by the banking systems of Western Europe and the United States; they were able to interpose Western commercial banks as a buffer between themselves and the 'high-risk' borrowers of Latin America and elsewhere. Latin American and some other less developed countries benefited. They were receiving large inflows of capital which in the mid-1970s, with high world inflation and low interest rates, were very cheap in real terms. Industrial nations also benefited because increased demand for capital equipment and machinery came from the borrowing nations.

Finally, the banks were content with the arrangements. Their assets, turnover and profits rapidly increased, and they developed new operational techniques which diminished their risks. First, for large loans with long maturities, the type of credit required by Latin American countries, they devised the roll-over credit. Although the loan to the Latin American country may have a long maturity (e.g. ten years), the interest rate is changed every time the credit is rolled over — usually every three or six months. This floating interest rate effectively passes to the borrower the risks of the market for it is the borrower who must cope with the long-term changes in the market. The second technique developed by the banking fraternity was the syndicated loan. Syndicated loans are simply credits shared by a large number of banks. In this way, the risks of default on large loans are spread very thin.

Anthony Sampson in his book, *The Moneylenders*,[11] argues that

Table 1.4: Latin American Debt — 1983

Country	Total Debt ($ billion)	Population (millions — 1979)	Per Capita Dollar Debt
Argentina	39.0	27.3	1428
Bolivia	3.8	5.4	704
Brazil	89.0	116.5	764
Chile	17.2	10.9	1642
Costa Rica	4.0	2.2	1818
Ecuador	6.5	8.1	802
Mexico	83.0	65.5	1267
Nicaragua	2.8	2.6	1077
Peru	11.5	17.1	672
Uruguay	4.0	2.9	1379
Venezuela	32.0	14.5	2207

Source: *Latin American Weekly Report, July 1983*

the banking profession is a cautious, herd-like profession. Bankers constantly watch other bankers to see what risks are being taken. If certain risks are seen to become reasonable by the banking fraternity, a high risk changes into a low risk relatively quickly. Such a pattern of behaviour can be detected in bankers' attitudes to Latin American and other Third World countries. In the mid-1970s, formerly high-risk countries, such as Brazil, rapidly became low-risk, and syndicated loans poured in. But this change of attitude affected only a few countries. The great majority of the poorer countries of Africa and Asia received virtually no syndicated loans. In Latin America, countries like Honduras, Guatemala, Paraguay and Surinam were avoided by the bankers and indeed remained net creditors to the private international banks.[12] In 1983, it was calculated that just 21 countries accounted for 84 per cent of external debt owed to banks by borrowers in less developed countries. Eleven of these were located in Latin America (see Table 1.4 and Figure 1.2).

Many Latin American countries thus became part of an international financial system in which many countries seemed to benefit apart from those poorer countries that bankers collectively thought of as high risk. In Latin America, the majority of countries were candidates for significant lending as Table 1.4 demonstrates. However, in retrospect, the international system of the 1970s relied on two vital ingredients for its success — low real interest rates and a healthy growth in the value of exports from the borrowing countries.

International economic conditions began to change dramatically in 1979, the year which recorded the second major rise in OPEC oil prices. The reaction of the international banks was to continue the recycling of the surplus petrodollars. But the world of 1979/80 saw the application of monetarist economic principles in two key western countries, the USA and the UK. While the major objective of bringing down inflation to less than 10 per cent was achieved in both countries, rising unemployment and a reduction in the level of economic activity resulted. Further, when the Reagan administration reduced taxation without a concomitant reduction in expenditure, large public sector deficits were created which needed to be financed. With strict monetary control in force at the Federal Reserve Bank, the US budget deficit had to be financed by large-scale borrowing. Such borrowing substantially increased the world demand for dollars and as a result interest rates rose — up and beyond the 20 per cent level.[13] Because of the USA's influence in the world economy, high interest rates became virtually a world-wide phenomenon.

The net effect of such events in the world economy was to reduce the level of economic activity in the industrialised countries. This in turn caused the stagnation of world trade in 1981, followed by a two per cent decline in 1982. The knock-on effect in Latin America was a reduction in both the volume and value of exports. In this way, an international financial system that benefited many Latin American countries in the 1970s became an enormous burden in the 1980s. In the 1970s, big Latin American borrowers such as Brazil and Mexico had financed rapidly increasing capital inflows through a rapid expansion of exports alongside low interest rates. With exports constrained and interest rates high, the massive capital borrowing suddenly became transformed into a huge deadweight on the economy. By 1983, Brazil's total debt had rised to $89 billion and Mexico's to $83 bn. Other countries had built up similarly high debts (see Table 1.4). It is difficult to transfer simple debt figures into an understanding of the burden they now impose on each country's economy. Other variables such as export performance, population totals and *per capita* income come into play. For example, if population is taken into account, an index of *per capita* debt can be calculated (see Table 1.4). Acording to this criterion, the big debtors in absolute terms (Brazil, Mexico) have economies less constrained than some smaller debtors such as Venezuela. If *per capita* debt is then compared to *per capita* income, an even

Figure 1.2: Latin American *Per Capita* Debt, 1983

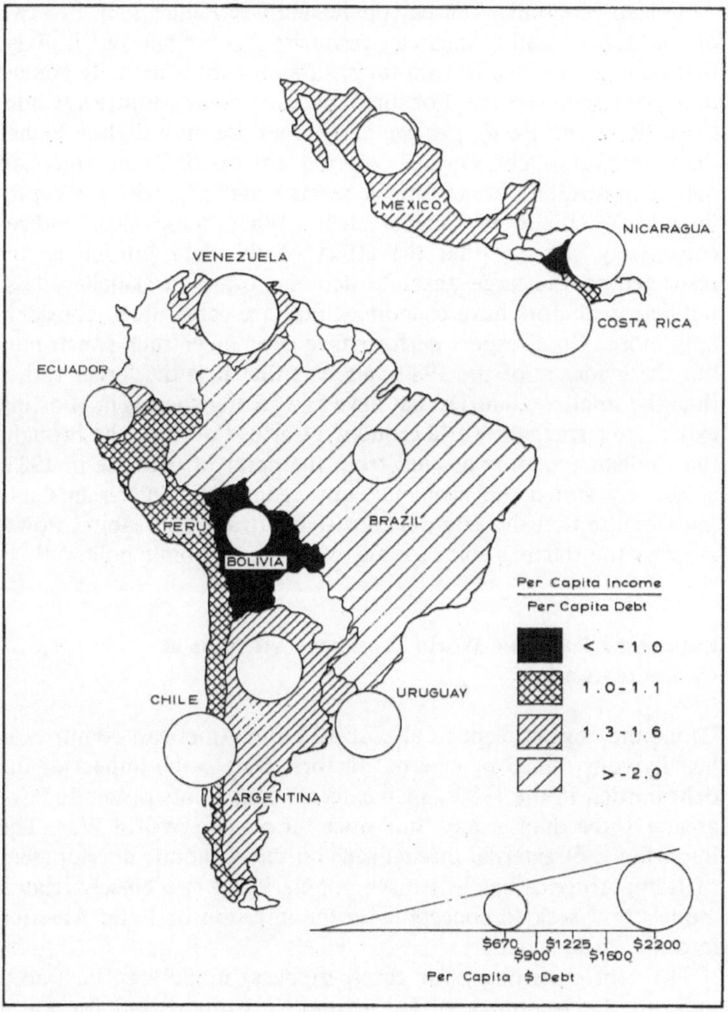

Source: Table 1.4

more dramatic indication of the burden of debt on some Latin American economies can be appreciated (see Figure 1.2). For two of the poorer Latin American countries (Nicaragua and Bolivia) that bankers decided to lend to, *per capita* debt is actually greater than *per capita* income. For three middle-income countries (Chile, Costa Rica and Peru), *per capita* incomes are only slightly higher than *per capita* debt. On this criterion, the major Latin American debtor in absolute terms, Brazil, seems better off, with *per capita* income 2½ times *per capita* debt. What these debt indices collectively show is that the effect of the debt burden is not restricted to the large absolute debtors; often the smaller, less-publicised debtors have economies that are constrained considerably more. Good export performance can relieve such constraints but the evidence of the 1980s suggests that it is the larger rather than the smaller countries that have had greater success in boosting exports in a stagnant world economy. Carlos Caceres, who brought the Chilean economy around from the point of collapse in 1983, effectively stated the problem; 'any economic minister in Chile must realise that the external constraints [trade and capital flows] are now the starting-point for any effective economic policy'.[14]

Latin America in the World Economy: Attempts at Conceptualisation

Economic development in almost all Latin American countries is heavily constrained by external factors. Due to the impact of the debt burden in the 1980s, such external constraints presently have greater force than at any time since the Second World War. The importance of external mechanisms on the economic development of Latin America has led to the popularity of two closely-related models that seek to conceptualise the insertion of Latin America into the world economy.

The centre-periphery (or core-periphery) model can be traced back to the beginning of the Economic Commission for Latin America (ECLA) and the writings of Raul Prebisch (ECLA's first director) in the late 1940s on the relations between Latin America and the industrialised economies.[15] Prebisch pointed out that the distinction between the centre (the industrialised economies) and the periphery (of which Latin America was part) found its inspiration chiefly in the unequal role of the two spatial sectors in the

Table 1.5: World Distribution of Manufacturing Output and Trade by Region, 1955 and 1975

Region	Distribution of manufacturing output and trade (percentages)					
	Output		Exports		Imports	
	1955	1975	1955	1975	1955	1975
North America	44.9	26.0	24.4	17.0	15.4	14.7
Western Europe	30.1	26.1	58.0	56.8	35.1	43.2
Eastern Europe/USSR	13.5	29.1	7.9	8.7	7.6	11.0
Japan	2.1	7.2	4.2	10.4	0.7	1.7
S. Africa/Australia/ New Zealand/Israel	1.8	1.7	0.9	0.8	7.0	3.3
Latin America	4.8	6.1	0.8	1.3	13.0	7.2
Africa	0.7	0.8	0.6	0.3	8.6	6.0
Asia	2.2	3.0	3.2	4.7	12.6	12.9

Source: UNECLA, *Analysis and Prospects of Latin American Industrial Development* (Second Latin America Conference on industrialisation, Cali, Colombia, 1979)

cyclical fluctuations of the world economy. Basically, the centre was active in times of recession (or growth) while the periphery was passive or reflexive.

In the original hypothesis at ECLA, the centre-periphery model was put forward from the point of view of trading relationships — the specialisation in primary product exports in Latin America and the specialisation of manufactured exports from the industrialised countries. Since the 1940s, the trading relationships between the centre and periphery have, as we have seen, become more complex but the basic pattern of centre and periphery still holds in terms of manufacturing output and trade. Table 1.5 demonstrates the asymmetry of Latin American manufacturing trade (in common with Asia and Africa) with exports of manufactures being low and imports high. Meanwhile, the countries of the centre enjoy the reverse with manufactured exports much higher than imports. Western Europe is defined as the major trading zone although Japan has the most extreme export-dominated trade asymmetry. The precise imbalance of manufacturing trade between Latin America and the core countries of the world can be illustrated by the 1975 trading figures. In that year, the core countries exported $30.8 billion of manufactured goods to Latin America and received only $3.7 billion worth in return — an unfavourable ratio of 8.1

for Latin America.

The centre-periphery model in its original formulation was rather restricted in scope, concentrating very much on the trading relationships between centre and periphery and in particular the terms of trade. Prebisch argued that the price relation had turned steadily against primary production from the 1870s until the Second World War and that this deterioration continued in the late 1940s and 1950s. The centre-periphery model was used to recommend a policy of rapid industrialisation in Latin America. Prebisch argued that only through rapid industrial growth and a concomitant expansion of manufactured exports would the adverse terms of trade be rectified between the centre of the world economy and the Latin American periphery.[16]

However, in the 1950s and 1960s, the centre-periphery model underwent considerable criticism and modification. Two ECLA economists, Pinto and Knakal, reviewed the centre-periphery model twenty years later and argued that it had been affected by 'two parallel and contradictory processes which could be called relative marginalisation and dependent insertion'.[17] Relative marginalisation of the periphery referred to the latter's declining share of world trade and especially its declining share of primary product trade. The major feature of post-Second World War trade has been the growth in trade between the industrialised countries of the centre itself. In this way, Pinto and Knakal saw the Latin American periphery as decreasingly 'necessary' for the centre in terms of providing primary products and a market for goods and services.

The other process envisaged by ECLA economists and taken up by a wide range of other social scientists was that of the dependent insertion of Latin American countries in the world economy. This concept of dependency came to have a much wider context than that of trade. It referred rather to a broader set of structural relations between Latin America and other peripheral countries, on the one hand, and the countries of the industrialised centre on the other. Peripheral countries were seen to be dependent on the industrialised countries in a wider context than that of trade. In terms of economic dependency, the processes of technology transfer, capital flows, indebtedness of peripheral countries and direct investment by multinational corporations were seen as further examples of dependency relationships. Dependency also became associated with social and cultural processes.[18]

The process of technology transfer and the question of foreign ownership and investment in Latin American development will be analysed in greater detail in later chapters. From the previous analysis of international trade and capital flows, one can make three points about the present validity and applicability of the centre-periphery and dependency models, nearly forty years after their inception.

First, the true nature of the dependent relationship is most obvious and acute during times of recession. As already outlined, the problems of high interest rates and stagnant exports from Latin American countries between 1980 and 1983 were both closely tied to the policies of the Reagan administration and other Western governments. The 1983/84 recovery of the industrialised nations, dominated largely by the United States, has taken place on terms that have increased rather than decreased the problems for the dependent nations. Economic growth of 6 per cent was recorded in the USA between mid-1983 and mid-1984. However, with tight US monetary control, a high budget deficit and a high dollar, the increasing growth in the USA was transmitted to the rest of the world as increased opportunities for exports to the USA, especially of manufactured goods. A high dollar alongside the traditionally high labour costs of the USA gave considerable opportunities for other countries with lower labour costs and weaker currencies to export their manufactures there. However, such benefits were mainly transmitted to other countries of the industrialised centre (Japan, UK, West Germany) and the newly-industrialising countries (NICs) — Taiwan, Singapore, Hong Kong, South Korea and, in Latin America, Brazil. Brazil managed to increase its exports in both 1983 and 1984 in order to achieve surpluses of $6.5 billion and an estimated $10 billion in the respective years. Such is the industrial strength and diversity of Brazil that it has been able to respond successfully to this favourable impetus to international growth. The majority of other Latin American countries have not been able to respond so favourably to the opportunities presented by the expanding North American market in 1983 and 1984. Chilean exports, for example, have remained stationary in terms of value from 1981 to 1984, despite a committed macroeconomic policy to the promotion of exports.

Instead most Latin American countries have been primarily affected by the more unfavourable side of recent recovery in the developed world. With the high US dollar sucking in exports, a large trade deficit has developed in the USA alongside its budget deficit. Such an unprecedented combination has caused the strongest currency in the world to become relatively scarce and for dollar interest rates to be kept high. As most Latin American debt is in dollars, the reverse side of the present recovery is the continuation of high interest rates on the large outstanding debts. For many Latin American countries, then, the recent recovery in the countries of the centre has not substantially changed the features of the 1979–82 recession; high interest rates persist alongside stagnant exports. In the 1979–82 recession and its aftermath, the true essence of the centre-periphery model is demonstrated — the centre active, the periphery passive and reflexive.

The second point about the present validity of the centre-periphery model must therefore be that the most acute external constraint on the development of the periphery is no longer trade but capital flow. As many countries in Latin America cannot even pay off the annual interest on their debt, indebtedness is increasing without any *actual* new capital arriving in the respective countries. Yet if a default is called, all forms of international capital would dry up, and the defaulting country would have trading problems to add to its financial problems.

Thirdly, trading relationships do still favour the industrialised countries of the centre, due to the continued decline in prices for primary products. There is an important differentiation here among the countries of the periphery, with those countries that are switching from primary product export to the export of manufactures performing better than those countries still heavily reliant on the export of unprocessed primary commodities.

The above analysis of the centre-periphery model, international trade and capital flows fully demonstrates that the economic development of Latin American countries is heavily constrained by external mechanisms. In this way, the capacity for independent manoeuvre in Latin American economies is much less than that of industrialised countries. However, I would also argue that any improvement in the manoeuvrability of Latin American economies must be closely linked to the process of industrialisation *within* those economies. Industrialisation must take place within a context of external constraints, but a successful process of industrialisation can effectively reduce these constraints. The following chapters on industrialisation in Latin America will present analyses within this framework.

References

1. R. Ffrench-Davis and E. Tironi, 1982, *Latin America and the New International Economic Order*. (Macmillan, London)

2. World Bank, 1983, *World Development Report* 1983. (Oxford University Press, New York)

3. R. Ffrench-Davis and E. Tironi, *Latin America and the N.I.E.O.*

4. A. Wood, 1984, When the bottom goes out of copper, *Geographical Magazine*, 56, 1, pp. 16–20

5. M. P. Todaro, 1977, *Economic Development in the Third World*. (Longman, London)

6. G. Perry, 1982, World Markets for Manufactures and Industrialisation in Developing Countries, in R. Ffrench-Davis and E. Tironi (eds.), *Latin America and the N.I.E.O.*

7. Ibid.

8. S. Griffith-Jones, 1982, Transnational Finance and Latin American National Development, *IDS Discussion Paper 175*, Sussex

9. Ibid.

10. R. N. Gwynne and S. Cunningham, 1983, The greatest debtors in the world, *Geographical Magazine* 55, 11, pp. 569–72

11. A. Sampson, 1981, *The Moneylenders*. (Hodder & Stoughton, London)

12. S. Griffith-Jones, Transnational Finance and Latin American Development

13. R. N. Gwynne, 1983, When trade stops being the engine of growth, *Geographical Magazine*, 55, 10, pp. 503–7

14. El Mercurio, January 25th, 1984

15. W. Baer, 1962, The economics of Prebisch and ECLA, *Economic Development and Cultural Change*, 10, pp. 169–82

16. Ibid.

17. A. Pinto and J. Knakal, 1973, The Centre-Periphery system twenty years later, *Social and Economic Studies*, 22, 1, pp. 34–89

18. For a survey of the political, social and political range of dependency theory, see G. Palma, Dependency: a formal theory of underdevelopment or a methodology for the analysis of concrete situations of underdevelopment? *World Development*, 1978, 6, 7/8, pp. 881–924

2 INDUSTRIAL POLICIES

Historical Perspectives

Arguments about the policies countries should adopt in order to industrialise go back to the late eighteenth century. Adam Smith stressed the need for countries to specialise in those commodities or industries in which they held a comparative advantage.[1] Hamilton in his 1791 *Report on Manufactures* noted, however, that 'the importation of manufactured supplies seem invariably to drain the merely agricultural people of their wealth' and that 'the United States cannot exchange with Europe on equal terms'.[2] The increasing demand for manufactured goods by the United States with economic expansion and a decreasing demand in Europe for primary goods would, he argued, expose the United States 'to a state of impoverishment'. He therefore recommended the protection of United States industry against European imports. The adoption of his policy by successive US governments paved the way for the rapid industrialisation of that country in the nineteenth century.

Latin American countries did not generally follow the example of their northern neighbour. Indeed from their own independence to 1930, they were basically a set of free trading countries, supplying a wide variety of agricultural and mineral raw materials to Europe in return for a wide variety of manufactured goods. There were examples of industrialisation linked to the growth in exports of primary products such as nitrates in Chile, coffee in Brazil and beef cattle in Argentina. In Argentina and Mexico, the industrial sector contributed 18 per cent and 14 per cent respectively to the domestic product by 1905. Dean demonstrated that the timing of São Paulo's industrial growth before 1930 was closely synchronised with good coffee exports, notably in the period between 1907 and 1913.[3] Furtado attempted to explain this link between early industrialisation and export growth as follows:

> Whereas the classical industrialisation experience was the result of innovations in productive processes which, by cutting prices,

> made possible substitution for craft manufacture and a creation of a home market for their own products, in the Latin American case the market was created as a result of the rise in productivity brought about by export specialisation and was at first supplied by imports.[4]

In this way, when the home market reached a certain size, a protectionist policy was enough to spark off an industrial upsurge. Moderate policies of protection were followed by some governments during the early twentieth century. Furthermore, the friction of distance gave industrialists in Latin America advantages in the manufacture of bulky products for the home market. Meanwhile, the First World War, by making industrial imports more scarce, also stimulated significant increases in manufacturing activity.

However, it was not until the world depression of the early 1930s that government became closely involved in the process of industrialisation. The initial involvement of government was the result of balance of payments crises caused by the precipitate decline of Latin American exports from an average of about $5,000 million in 1928–9 to $1,500 million in 1933. With the concomitant decline in foreign exchange, Latin American governments found themselves able to finance decreasing amounts of imported manufactured goods from the industrial countries. As a result various measures were taken to conserve and ration decreased foreign exchange resources. Tariffs were raised by most countries, import quotas were enforced and restrictions placed on the use of foreign exchange. An array of policies was developed out of the crisis, and administrative machinery established to carry them out. In general terms, Latin America changed from a set of free-trade economies to one of highly protected economies. Tariffs, quotas, and exchange controls provided protection from foreign competitors by making the entry of foreign goods expensive or impossible. Latin American entrepreneurs, observing the scarcity of goods and the level of protection, began to produce or increase the production of goods previously imported. Industrial production and employment increased as a result. In Chile, industrial employment increased by over 5 per cent per annum between 1928 and 1937, but in those sectors which protection particularly favoured employment rose at much higher rates — 10 per cent per annum in metal products, 16 per cent in textiles and 19 per cent in chemicals.[5]

Such a strategy of industrial development behind high protective tariffs continued to be followed in most Latin American countries after the adverse effects of the depression had diminished. It was noted that all major industrial countries had industrialised behind high protective tariffs. In was only after a country had developed a mature industrial structure that it could become involved in the free trading of industrial goods. Protective policies also promoted the development of infant industries, so that national firms could learn new technologies and new processes without too much fear of making large losses. Protection towards industry was also favoured for reasons of employment. Industrial goods that are produced nationally employ labour whereas imported goods obviously do not. Furthermore, employment in manufacturing had both higher productivity and wages than other sectors of the economy so that protective policies could radically change the labour market and increase the national product.

The Policy of Import Substitution

Because of such perceived advantages of protection for industrial development, protective policies were maintained after the Second World War and in the 1950s became formalised into the policy known as 'import substitution'. Adherents of the policy came to envisage four stages of industrial production within this process of industrialisation. The first stage saw the production of basic non-durable consumer goods such as textiles, foodstuffs and pharmaceuticals. This was followed by the production of consumer durable products, such as cookers, radios, televisions, and the critical motor vehicle industry; assembly began with a considerable ratio of imported parts. The third stage was critical in the industrialising process as it had to promote 'intermediate' industries producing the inputs for companies set up during the first and second stages. Typical industries at this stage would be chemical plants making paint, synthetic fibres, dyes and acids, or engineering works producing small motors and gearboxes, or parts industries for durable goods assembly. The final stage of the process would promote the development of the capital goods industry which would manufacture machinery and plant installations. It was the task of the government to plan and synchronise each subsequent stage in the process.

However, as import substitution industrialisation progressed through the 1950s and 1960s, it became evident that despite its theoretical attraction, high-cost industry normally resulted, particularly in the smaller countries where market size was limited. A United Nations study that analysed eleven non-durable and nine durable consumer good sectors in 1964 concluded that 'the prevailing situation in the region is characterised by high relative prices of manufactured products and this phenomenon cannot but affect the size of Latin America's market for the type of good'.[6]

High costs and high prices behind high protective tariffs were due to a wide variety of factors, but one key structural problem was the lack of economies of scale that small markets permitted, particularly in the critical second and third stages of the import substitution process. As production increases in any operation, unit costs will normally decline as fixed costs (e.g. plant and technology) will be spread over more and more units. The decline in unit costs with increased production will slacken and even out at some stage (known as the minimum efficient scale) before perhaps rising as output increases still further. Figure 2.1 shows the notional relationship between output of a vehicle assembly operation and unit costs in Latin America. The minimum efficient scale is reached at approximately 200,000 vehicles per annum but there is a low per unit cost penalty for annual output between 100,000 and 200,000.

The example of the motor vehicle industry can demonstrate the relationship between economies of scale and actual production levels in Latin America.[7] Table 2.1 shows vehicle output in each producing country by firm for 1977. It is evident that the assembly plant economies illustrated in Figure 2.1 are only achieved by firms operating in Brazil (Volkswagen, General Motors, Ford and Fiat). Firms in other countries incurred severe cost penalties in their assembly operations. As a result, high-cost cars were being produced and sold to national consumers. A detailed cost analysis of the Chilean motor vehicle industry in 1969 demonstrated that cars were produced at up to three times the cost in the vehicle's country of origin.[8] Cars in both Argentina and Venezuela were being sold at double international levels in the late 1970s.[9]

Such high prices and high costs were not solely the result of a lack of sufficient economies of scale in the assembly of vehicles. In order to promote certain stage three industries, many governments instituted schemes whereby assembly firms had to purchase increasing quantities of nationally-produced components. However, components production can require greater production runs than

Table 2.1: Distribution of Motor Vehicle Production by Country and Firm in Latin America, 1977

	Argentina	Brazil	Chile	Colombia	Mexico	Peru	Venezuela	TOTAL	%
Chrysler	23,434	21,970	–	8,275	57,956	7,169	32,430	151,234	9.1
Fiat	47,837	77,963	3,120	4,128	–	–	4,889	137,937	8.3
Ford	56,795	130,197	–	–	50,503	–	61,665	299,160	17.9
General Motors	20,897	153,836	2,124	–	34,638	–	32,281	243,776	14.6
Mercedes Benz	7,845	52,957	–	–	–	–	564	61,366	3.7
Nissan	–	–	–	–	37,066	5,270	5,223	47,559	2.9
Peugeot/Citroen	34,214	–	5,277	–	–	–	–	39,491	2.4
Renault	34,744	–	2,568	17,353	27,559	–	5,153	87,377	5.2
Volkswagen	–	472,192	–	–	52,143	6,075	5,340	535,750	32.1
Others	9,590	10,127	–	–	20,948	6,710	15,752	63,127	3.8
TOTAL	235,356	919,242	13,089	29,756	280,813	25,224	163,297	1,666,777	100.0
Percentage	14.1	55.2	0.8	1.8	16.8	1.5	9.8	100.0	

Source: R. N. Gwynne, The Motor Vehicle Industry in Latin America, *Bank of London and South America Review*, 1978, 12(9), 462–71

Figure 2.1: Notional Relationship between Output of a Vehicle Assembly Operation and Unit Costs in Latin America

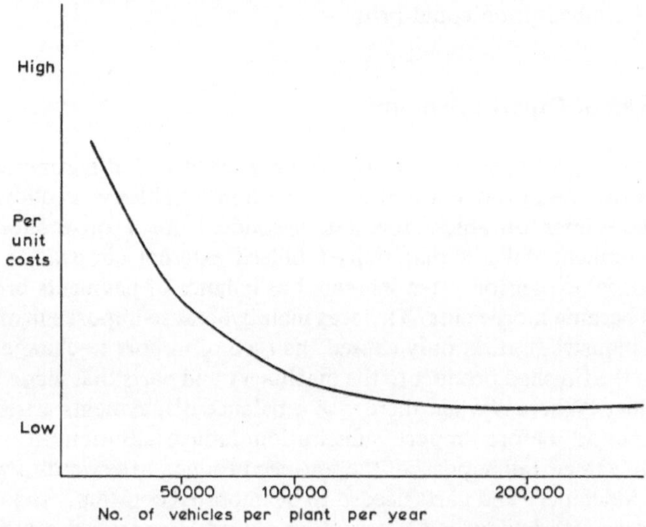

vehicle assembly in order to achieve maximum economies of scale. Modern engine and transmission plants attain maximum economies of scale at annual production levels of 500,000 per annum (although cost penalties are not too great once 100,000 per annum levels are reached). As only the Brazilian assembly industry could offer such a level of demand, the Latin American motor vehicle industry has often been characterised by small-scale plants assembling high-cost components for a domestic market limited by the high prices of the finished product.

Developing in the late 1950s and 1960s, the motor vehicle industry was critical to the process of import substitution industrialisation. As an assembly industry, it produced one of the more sophisticated goods of the second stage. However, in terms of government planning, its critical role was to develop a variety of components industries which, through backward linkage, would create additional demand for the national production of such basic goods as steel, alloys, aluminium, plastics and glass. Its eventual high cost caused scepticism about the applicability of import substitution for the efficient production of more sophisticated products and for the ability of the process to advance further than stage two without the aid of very high tariffs. Nevertheless, in this context, such factors as market size and character became crucial as large, rapidly growing markets seemed to provide conditions in which import substitution could prosper.

Policies of Export Promotion

As the import substitution process progressed in Latin American countries, and production of consumer non-durables was followed by consumer durables, the less dependent form of economic development did not materialise. Indeed external constraints on industrial expansion often increased as balance of payments problems became more acute. This was mainly because import substitution industrialisation only caused the *type* of import to change — from the finished product to the machinery and parts that made the product. Whereas when there was a balance of payments crisis in the period before import substitution industrialisation, it was feasible to curtail imports of the finished product, afterwards it was not. Machinery and parts needed to be imported constantly to keep factories producing and workers employed. Import substitution industrialisation thus had the reverse effect of that anticipated. It increased the dependence of individual countries on the world trading system and decreased their room for manoeuvre in balance of payments crises and recessions.

In order to have more room for manoeuvre and to reduce the external constraints on their development some Latin American countries adopted a more outward-looking approach to industry in the 1970s. The exploitation of comparative advantage and the promotion of exports were dominant characteristics of this new approach. Small countries wished to overcome the problems of the small size of their market for manufactured goods. Larger countries wished to develop successful export sectors from industries previously only serving the domestic market.

Brazil has been very much in the forefront of reorientating the focus of industrial policy. The reorientation has been gradual, accelerating rapidly in the acute domestic recession of 1980 to 1984. The Brazilian policy has been one of maintaining as protected a market as possible whilst stimulating industrial exports by giving manufacturers an effective subsidy through a tax rebate. In 1981 this tax rebate (*Imposto sobre Produtos Industrializados*) was

Figure 2.2: Brazilian Exports, 1971–81

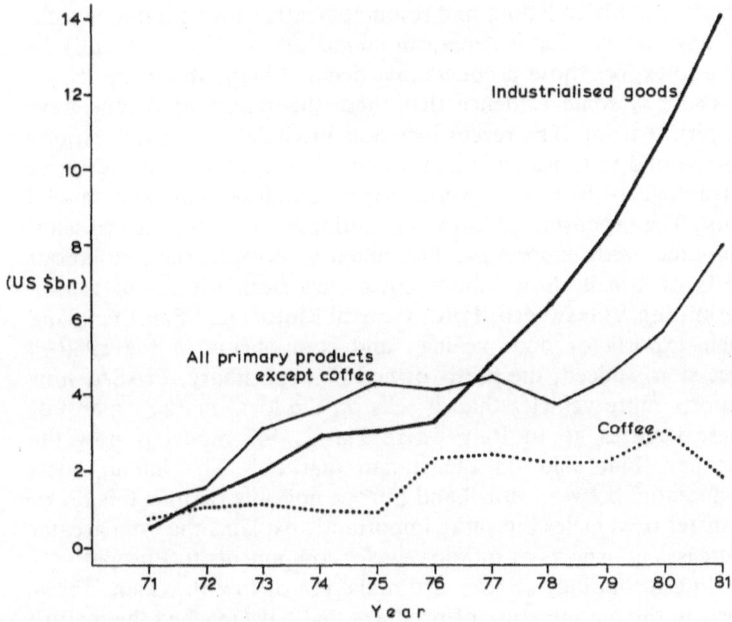

Source: Banco do Brasil

equivalent to a 15 per cent subsidy on export sales, although it was reduced to 9 per cent in 1983. As a result, the export of industrial goods (both manufactured and semi-manufactured) which constituted only about 10 per cent of total exports in 1971, now accounts for as much as 50 per cent (see Figure 2.2). In the 1980–1 domestic recession, exports of chemicals increased in value by over 100 per cent, black and white T.V. sets by 70 per cent, shoes and cutlery by about 50 per cent and motor vehicles by around 40 per cent. Motor vehicle and components exports now account for almost 10 per cent of Brazil's total exports, a fact that has made it a leading vehicle exporter, more important than the United States.

The advantages of an industrial policy in which the promotion of exports is a vital component is that the country can reduce the internal and domestic constraints on industrialisation by concentrating on the production either of goods in which it has a comparative advantage or those which use factors of production with which it is particularly favoured. In Latin America, the latter applies mainly to labour and resources (rather than capital or technology) so that Latin American countries should theoretically be able to export those products that demand high labour inputs.

There is some evidence that these theoretical assertions have empirical truth. The recent increases in Chile's industrial exports correspond to those sectors in which it has a comparative resource advantage — fish meal, wood, paper, cellulose, wine and canned fruit. The expansion of Brazil's manufactured (as opposed to semi-manufactured) exports has had much to do with the low labour costs of Brazil. Low labour costs have been the major reason permitting Volkswagen, Ford, General Motors and Fiat to expand their exports of both vehicles and components in the 1980–4 recession. Indeed, the partly-owned Fiat subsidiary, FIASA, now exports more vehicles than it sells on the local market; many of these vehicles go to Italy where Fiasa's 147 model is now the cheapest Fiat sold on the Italian market.[10] The labour costs differential between Brazil and Europe not only permits this direct transfer of vehicles but more importantly explains the even greater increases in exports of motor vehicles components to Europe.

High technology sectors also make use of low Brazilian labour costs in the manufacture of products that have reached the mature phase in their product cycle. Kuznets's concept of the product life cycle is being increasingly applied to worldwide shifts in manufacturing production.[11] According to Kuznets, the life of a product can be divided into three phases (see Figure 2.3). The 'new' phase after the invention of the product is a period of low sales and low sales growth as the manufacturing and marketing problems are gradually solved. In terms of location, this process of 'putting the product on the market' is increasingly concentrated in the USA due to the latter's pool of entrepreneurial talent and large high-income market.[12] In the 'growth' phase, the product becomes cheaper, appeals to wider sections of the population both in the USA and outside, and manufacturing takes place in other countries of the centre — Japan, West Germany, UK, France. The 'growth' phase constitutes an interlinked spiral of decreasing unit costs, increasing demand and expanding productive capacity. By the end of the 'growth' phase, systems of mass production and marketing have been organised and economies of scale taken advantage of. With production, machinery and marketing standardised, the 'mature'

Figure 2.3: Stages in the Product Life Cycle

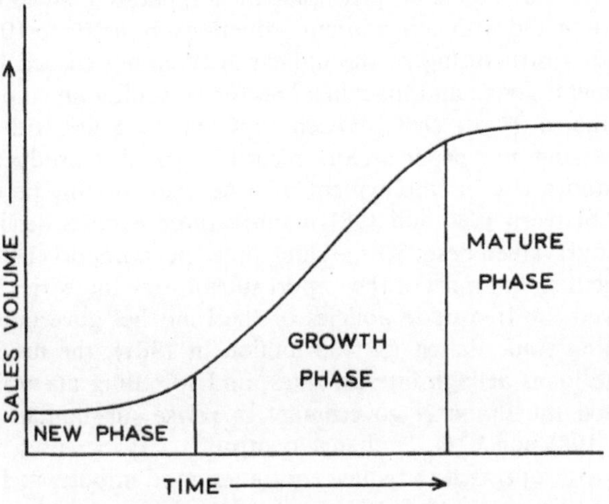

Source: L. G. Franko, *The European Multinationals* (Harper & Row, London 1976)

phase begins — a period of high sales but little growth as markets are relatively saturated. In order to reduce costs further, world production shifts to the less developed countries; standardised production means few 'teething' problems and lower labour cost reduces the cost of the final product. In Brazil, there is now a wide variety of 'mature' products being made for export to foreign markets. Some of these products are controlled by foreign firms, some by Brazilian. The Japanese camera company, Yashica, now makes nearly 200,000 cameras a year in Brazil of which 60 per cent are for export.[13] The Brazilian hi-fi makers, Gradiente and Sharp do Brasil, export significant amounts of standardised hi-fi equipment. In 1979, Gradiente took over the British hi-fi firm, Garrard, only to close the latter's UK plant in 1982; at that time labour costs in the UK were four times those of the Brazilian plant.[14] The product life cycle applies mainly to locational shifts in those industries in which there are well-defined labour-intensive stages of production.

However, the most radical reorientation of industrial policy in terms of the shift from import substitution to export promotion occurred in Chile after 1974 under the Pinochet government. Chile left the Andean Pact in 1976 in order to pursue unhindered economic policies based on principles of comparative advantage and free trade and with concomitant reductions in tariffs to 10 per cent. Major restructuring of the industrial economy ensued with the large metal goods and machinery sector recording an employment decline of 28 per cent between 1974 and 1978 but with the food processing and paper sectors recording sizeable production increases and a rise in employment of 6 per cent for the period. However, between 1980 and 1981 manufactured exports declined due to an overvalued peso. At the same time cheap imports caused grave difficulties for most of the import substituting industries that had survived the free trade policies of the Pinochet government. The resulting trade deficit ($2,458 million in 1981), the need to repay large loans at high interest rates, and spiralling unemployment caused the Pinochet government to revise substantially its policies in 1983 and 1984. Exchange controls, tariff increases, and exchange taxes operated to reduce consumer good imports in 1983 to about one quarter of their 1981 level although industrial exports remained stagnant. With the formal expulsion of the monetarist 'Chicago boys' from the Chilean Treasury in April 1984, it would appear that export-led industrial growth has been deemed a failure and a return to a more protected market has been engineered. Between 1983 and 1984, industrial production increased by 12 per cent, with eight import-substituting sectors recording growth rates in excess of 25 per cent — china, machinery, furniture, electrical products, tyres, textiles, clothing and glass (see Chapter 11).

Other Latin American countries have attempted to reorientate their industrial policies towards exports. Colombia and Peru have been following the Chilean example and have generally reduced tariffs to the 30 per cent level. The oil-exporting countries of Mexico and Venezuela have attempted to promote manufactured exports by subsidies, but high protection rates, overvalued currencies and an expanding domestic market have meant that most industrialists have not shifted their gaze from the national horizon.

In the last twenty years, export promotion policies have become more widely accepted in Latin America than policies of import substitution. Countries that have shifted gradually from one policy to the other seem to have had the best results. Countries that have kept firmly to policies of import substitution have generally had stagnant manufacturing sectors during the last ten years. Venezuelan manufacturing's share of national GDP fell from 16.2 per cent in 1972 to 15.8 per cent in 1982.[15] Productivity in Venezuelan manufacturing actually fell by 5 per cent between 1974 and 1977.[16] Meanwhile, countries that have shifted radically and drastically from one policy to another have not performed well overall. Chile's manufacturing GDP in 1983 was lower than that of 1974 after the 1981–3 decline in manufacturing production.[17] It is worth quoting Paul Streeten's comment on over-zealous export promoters:

> It is sometimes argued that the pressures for efficiency are greater in an open economy with substantially reduced protection and "realistic" exchange rates. The *general* competitive pressures are of course stronger in such an economy, although it must be remembered that the cold winds of competition can wither, as well as strengthen, tender plants.[18]

The process of industrialisation and policy change in Latin America reflects the differences between large and small countries as they attempt to industrialise. Chenery and other World Bank economists have argued that as countries now attempt the transition in productive structure from primary production to manufacturing industry, they must recognise 'the existence of the advanced countries as sources of technology, capital and manufactured imports, as well as markets for exports'.[19] The importance of economies of scale therefore becomes central to the process of industrialisation in the present industrialising countries. For larger countries, this means that at low levels of *per capita* income, they can promote a substantial industrialising process orientated towards supplying the domestic market. However, once moderate levels of *per capita* income are achieved, Chenery argues that larger countries need to export in order to carry on the process of industrial growth. However, a very different strategy of transitional growth is recommended for smaller countries. From low levels of *per capita* income, small countries need to concentrate on developing certain specialised sectors in manufacturing in order to promote exports. Poor, small countries should concentrate initially on the production and export of such goods as textiles, shoes and clothing. As *per capita* income increases, production and exports should gradually diversify. In this way, increased *per capita* income is closely linked to the process of diversification and export growth in the manufacturing sector. The importance of trade in manufactured goods for the process of industrial growth is borne out by the Latin American experience previously discussed. Larger countries can delay the process longer than smaller countries, but it is still necessary for a shift from import substitution to export promotion to take place.

References

1. A. Smith, 1970, *The Wealth of Nations.* (Penguin, Harmondsworth)

2. J. Grunwald, 1970, Some reflections on Latin American industrialisation policy, *Journal of Political Economy*, 78, pp. 825–56

3. W. Dean, 1969, *The Industrialization of São Paulo, 1880–1945.* (University of Texas Press, Austin)

4. C. Furtado, 1970, *Economic Development of Latin America.* (Cambridge University Press, Cambridge)

5. R. N. Gwynne, 1978, *Industrial Decentralisation in Chile.* (Unpublished doctoral dissertation, University of Liverpool)

6. United Nations Economic Commission for Latin America (UNECLA), 1979, *International Cooperation for industrial development in Latin America.* (Second Latin American Conference on Industrialization, Cali, Colombia)

7. See R. N. Gwynne, 1978, The Motor Vehicle Industry in Latin America, *Bank of London and South America Review*, 12(9), pp. 462–71

8. Instituto de Costos, 1969, *Estudio sobre la industria automotriz.* (Santiago)

9. R. N. Gwynne, 1979, The Venezuelan automobile industry, *Business Venezuela*, 64, pp. 24–28

10. *Latin American Weekly Report*, 1982, No. 11, 23 March

11. R. Vernon, 1966, International investment and international trade in the product cycle, *Quarterly Journal of Economics*, 80, pp. 190–207, also L. T. Wells (ed.), 1972, *The Product Life Cycle and International Trade.* (Harvard University Press, Boston)

12. L. G. Franko, 1976, *The European Multinationals.* (Harper & Row, London)

13. *Latin America*, 1981, Regional Report, Brazil, September

14. *Latin America Weekly Report*, 1983 January 21st

15. Ministerio de Energia y Minas, 1983, *Petroleo y Otros Datos Estadisticos 1982.* (Caracas)

16. R. N. Gwynne, 1979, The plight of productivity, *Business Venezuela*, 63, pp. 27–30

17. E. Errazuriz, 1984, La desindistrialización del país, *Mensaje*, 33, 326, pp. 41–5

18. P. Streeten, 1982, A cool look at "outward-looking" strategies for development, *World Economy*, 5, 2, pp. 159–69

19. H. B. Chenery, 1977, Transitional Growth and World Industrialisation, in B. Ohlin et al (eds.), *The International Allocation of Economic Activity* (Macmillan, London and Basingstoke)

3 INDUSTRIAL STRUCTURE AND TECHNOLOGICAL CHANGE

Industrial Structure

In spite of the contrasting records and problems of import substitution and export promotion as policies of industrial growth, industrialisation has nevertheless proceeded apace in almost all Latin American countries since the Second World War. Overall GDP increased at an average annual rate of 6.5 per cent between 1950 and 1978 (see Table 3.1). Thus, there were striking increases in the 'level of industrialisation' (defined as the share of the manufacturing product in the total product) from 10 per cent in 1950 to 26 per cent in 1978 and in *per capita* manufacturing GDP — from $90 in 1950 to $240 in 1978 (both at 1970 prices). Significant spatial variations in the level of industrialisation had developed by 1978. The larger countries of Brazil, Mexico and Argentina had 30 per cent, 26 per cent and 33 per cent of their respective GDPs accounted for by manufacturing while the oil-exporting countries of Venezuela (17 per cent) and Ecuador (20 per cent) and the smaller countries of Central America (18 per cent) had much lower proportions (see Tables 3.2 and 3.3).

The manufacturing growth rates presented in Table 3.1 basically demonstrate the influence of import substitution and economic integration policies on industrial development;[1] the effects of the shift to more export-oriented policies in the 1970s and 1980s are less evident. As a result the large countries, where import substitution had the advantage of supplying large national markets, recorded much higher manufacturing growth rates than other countries, particularly in the 1965–73 period when average growth rates of 12 per cent in Brazil and 8 per cent in Mexico were achieved. The industrial achievements of the countries of the Central American Common Market are reflected in the 7.4 per cent average annual growth rate from 1950 to 1973, while the limited success of the Andean Group is demonstrated by the relatively low growth rate of 4.1 per cent between 1973 and 1978. Meanwhile, the failure of import substitution policies to provide industrial growth outside large countries is most clearly seen in the River Plate countries

Table 3.1: Growth Rates of Population and the Manufacturing Product in Latin America 1950–78

	Population	Manufacturing Product			
	1950–78	1950–65	1965–73	1973–8	1950–78
Large Countries	3.1	7.3	10.4	6.3	8.0
Brazil	3.0	7.3	12.0	6.3	8.5
Mexico	3.3	7.2	8.1	6.3	7.3
Andean Countries	3.0	6.8	5.6	4.1	6.0
Bolivia	2.4	2.0	5.6	8.3	4.2
Chile	2.1	5.5	3.4	−1.4	3.7
Colombia	3.2	6.2	7.7	5.4	6.5
Ecuador	3.2	5.3	7.1	11.9	7.0
Peru	2.8	7.8	6.6	1.8	6.4
Venezuela	3.5	9.5	5.0	7.6	7.9
Central American Countries	3.0	7.4	7.4	5.3	7.0
Costa Rica	3.2	7.9	9.4	8.1	8.3
El Salvador	3.1	7.2	5.8	5.2	6.4
Guatemala	2.8	5.4	7.7	6.2	6.2
Honduras	3.2	8.3	6.4	6.3	7.4
Nicaragua	3.0	9.5	6.9	3.6	7.6
Panama	2.9	10.1	8.1	0.2	7.7
River Plate Countries	1.6	4.6	5.6	−0.4	4.0
Argentina	1.6	4.8	5.9	−1.0	4.1
Paraguay	2.7	3.3	6.0	7.2	4.8
Uruguay	1.4	2.7	0.9	5.9	2.7
Latin America	2.8	6.3	8.2	4.5	6.5

Source: CEPAL

where a negative industrial growth rate was recorded between 1973 and 1978. Arguably, the import substitution process was 'completed' earlier in Argentina and Uruguay, but with the high costs of national industry preventing exports and further domestic expansion, such 'completion' brought industrial stagnation. Argentine industry declined by 1 per cent per annum between 1973 and 1978 causing *per capita* manufacturing GDP to fall from $531 in 1973 to $474 in 1978. In this way, the industrial weight of the River Plate region in Latin America fell from 35 per cent in 1950 to only 18 per cent in 1978. Meanwhile, the two larger countries that had forged a more competitive industrial structure from import substitution, Brazil and Mexico, had increased their combined

Table 3.2: Manufacturing and Other Indices for Latin American Countries, 1950

	Population (thousands of inhabitants)	GDP (millions of 1970 dollars)	Manufacturing GDP (millions of 1970 dollars)	Level of Industrialisation (Manufacturing GDP as % of GDP)	Industrial weight within the region (%)	*Per Capita* Manufacturing GDP (1970 prices)
Large Countries	79,507	27,683	5,597	20	42.5	70
Brazil	52,901	14,440	3,113	22	23.6	59
Mexico	26,606	13,243	2,484	19	18.9	93
Andean Countries	36,928	16,471	2,600	16	19.8	70
Bolivia	3,019	754	104	14	0.8	34
Chile	6,019	3,914	898	23	6.8	149
Colombia	11,689	4,658	585	13	4.5	50
Ecuador	3,224	867	137	16	1.0	42
Peru	7,832	2,774	454	16	3.5	58
Venezuela	5,145	3,504	422	12	3.2	82
Central American Countries	9,168	2,855	327	12	2.5	36
Costa Rica	866	335	49	15	0.4	58
El Salvador	1,940	554	76	14	0.6	39
Guatemala	3,054	947	114	12	0.9	37
Honduras	1,390	359	25	7	0.2	17
Nicaragua	1,109	261	30	11	0.2	27
Panama	809	399	33	8	0.2	41
River Plate Countries	20,715	18,270	4,640	25	35.2	224
Argentina	17,150	15,699	4,103	26	31.2	239
Paraguay	1,371	430	68	16	0.5	50
Uruguay	2,194	2,141	469	22	3.5	214
Total Latin America	146,318	65,279	13,164	20	100.0	90

Source: CEPAL

Table 3.3: Manufacturing and Other Indices for Latin American Countries, 1978

	Population (thousands of inhabitants)	GDP (millions of 1970 dollars)	Manufacturing GDP (millions of 1970 dollars)	Level of Industrialisation (Manufacturing GDP as % of GDP)	Industrial weight within the region (%)	*Per Capita* Manufacturing GDP (1970 prices)
Large Countries	184,898	170,140	48,172	28	62.4	261
Brazil	119,477	101,056	30,327	30	39.3	254
Mexico	65,421	69,084	17,845	26	23.1	273
Andean Countries	83,394	66,662	13,132	20	17.0	158
Bolivia	5,848	2,072	325	16	0.4	56
Chile	10,843	10,335	2,451	24	3.2	226
Colombia	28,424	19,162	3,384	18	4.4	119
Ecuador	7,798	4,434	905	20	1.2	116
Peru	17,148	10,323	2,554	25	3.3	149
Venezuela	13,333	20,336	3,513	17	4.5	264
Central American Countries	21,002	12,279	2,198	18	2.8	105
Costa Rica	2,111	2,031	461	23	0.6	218
El Salvador	4,524	2,238	436	19	0.6	96
Guatemala	6,623	3,783	617	16	0.8	93
Honduras	3,362	1,166	185	16	0.2	55
Nicaragua	2,559	1,195	238	20	0.3	93
Panama	1,823	1,866	261	14	0.3	143
River Plate Countries	32,490	43,042	13,761	32	17.8	424
Argentina	26,395	38,011	12,512	33	16.2	474
Paraguay	2,888	1,553	250	16	0.3	87
Uruguay	3,207	3,478	999	29	1.3	312
Total Latin America	321,784	292,123	77,263	26	100.0	240

Source: CEPAL

industrial weight in Latin America from 43 per cent to 62 per cent — a figure that effectively summarises the close links between size of country and industrial success in import substitution. Economic integration was not effective in breaking the rule. The industrial weighting of the twelve smaller countries involved in regional schemes of economic integration fell slightly from 22 per cent in 1950 to 20 per cent in 1978.

Import substitution industrialisation has brought major structural changes in manufacturing. As would be expected from the import substitution model, industries producing non-durable consumer goods have declined in relative importance as the industries producing intermediate goods, consumer durables and capital goods have developed. In 1950 non-durable consumer goods represented almost two thirds of total manufacturing production, as opposed to 40 per cent today. In contrast, the relative importance of intermediate products in manufacturing output has risen from less than 25 per cent of the total to more than one third at present. However, it is the consumer durables and investment goods sector where the relative change has been most striking. This sector represented 11 per cent of total manufactures in 1950 and now accounts for more than one quarter of manufactures produced.

Again, there are great differences among Latin American countries according to size of country. Table 3.4 demonstrates that the process of import substitution has been much more 'complete' in the large countries than in the smaller countries. Intermediate manufactures and consumer durable/capital goods sectors accounted for 37 per cent and 28 per cent respectively of total value-added in the larger countries (including Argentina) by 1975, against 35 per cent and 17 per cent in medium-sized countries and only 26 per cent and 9 per cent in small countries. Indeed in the period 1950–75, as much as 83 per cent and 87 per cent of Latin American growth in intermediate and consumer durable/capital goods respectively took place in the three larger countries of Brazil, Mexico and Argentina (see Table 3.5). The production of non-durable consumer goods has, however, expanded considerably in medium-sized and small countries, where 48 per cent and 65 per cent of total value-added still corresponds to these sectors and where 17 per cent and 12 per cent of Latin American growth in these goods took place.

The production of intermediate goods has expanded significantly in medium-sized countries but less so in small countries, although

Table 3.4: Industrial Structure and Size of Country in Latin America 1950–75 (percentage of value added in the manufacturing sector)

		Non-durable consumer manufactures	Intermediate manufactures				Consumer and capital manufactures	Total manufactures
		A	B	C	D	B+C+D	E	
Large countries	1950	64	10	10	4	24	12	100
	1975	35	9	21	7	37	28	100
Medium-sized countries	1950	66	10	15	3	28	6	100
	1975	48	9	20	6	35	17	100
Small countries	1950	85	8	6	—	14	1	100
	1975	65	11	14	1	26	9	100

A. Food, beverages and tobacco (Division 31); textile, wearing apparel and leather industries (Division 32); furniture and fixtures, except primarily of metal (major group 332); printing, publishing and allied industries (major group 342); pottery, china and earthenware (major group 361), and other manufacturing industries (Division 39), according to ISIC

B. Wood and cork products, except furniture (major group 331); paper and paper products (major group 341); glass and glass products (major group 362), and other non-metallic mineral products (major group 369), according to ISIC

C. Chemicals and chemical products, petroleum, coal, rubber and plastic products (Division 35 of ISIC)

D. Basic metal industries (Division 37 of ISIC)

E. Metal products, machinery and equipment (Division 38 of ISIC)

ISIC — International Standard Industrial Classification

Source: CEPAL

Table 3.5: Sectoral Growth and Size of Country in Latin America, 1950–75

	Total manufacturing industry	Manufactures[a]				
		Non-durable consumer goods	Intermediate goods			Consumer durables and capital goods
		A	B	C	D	E
	%	%	%	%	%	%
Large countries	80	71	80	83	86	87
Medium-sized countries	13	17	13	13	13	10
Small countries	7	12	7	4	1	3
Total Latin America	100	100	100	100	100	100

a. Columns A, B, C, D and E represent the ISIC manufacturing branches and groups as indicated at the foot of Table 3.4
Source: CEPAL

average growth rates were registered in Table 3.5's B classification of intermediate goods — paper, glass, wood and non-metallic mineral products; for example, in the cement industry, 30 per cent of Latin American production corresponded to medium-sized and small countries in 1976. However, growth in the D classification of intermediate goods (basic metals such as steel and aluminium) is almost entirely restricted to the larger countries and the one medium-sized country, Venezuela, that benefits from considerable advantages (in the form of cheap materials and power) in both steel and aluminium production at Ciudad Guayana. In 1976, 94 per cent and 100 per cent of Latin American steel and aluminium production respectively corresponded to Argentina, Brazil, Mexico and Venezuela.

The most pronounced concentration of industrial growth in the larger countries occurred in the consumer durable/capital goods sectors (87 per cent of continental growth). In terms of motor vehicle production, 86 per cent of assembly took place in Brazil, Mexico and Argentina in 1977. Furthermore, the governments of these countries have demanded high integration levels, normally the equivalent of over 90 per cent of parts by value from national sources. The medium-sized countries have been much less ambitious and accepted levels of 30 per cent or below. Meanwhile, the capital goods sector has 90 per cent of its production concentrated in the three largest countries, with Brazil having the most technologically advanced and broadest machinery sector; indeed, in certain areas, such as machinery and plant for sugar-cane mills and alcohol production, Brazilian firms compete successfully on the world market.

Technological Change

The concentration of the production of intermediate, consumer durable and capital goods in Brazil, Mexico and Argentina is intimately linked not only to size of market but also to the internal generation and adaptation of new technology. In this respect, it is worth noting the comments of Jorge Katz, head of the UNDP-IDB Research Project on Technology Transfer in Latin America. According to Katz 'Argentina, Brazil and Mexico have travelled far enough along the road of industrial modernisation and technological maturity to separate them from the rest' of Latin America.[2] These three larger countries have been able to promote the following in their relations with other Latin American countries:

1. The export of manufactured goods of significant technological complexity — vehicles, machinery, antibiotics, chemicals, electronic goods,
2. The sale of technology under licence,
3. Direct investment,
4. Technical assistance in the development of basic infrastructure, such as atomic energy and transport facilities.

Katz goes on to argue that these new technological and commercial manifestations reveal an incipient form of 'internationalisation' within the region, which was virtually unknown in Latin America until recently. Furthermore, he forecasts that 'the gradual strengthening of the international position of a few developing countries will certainly be the rule and not the exception in years to come'.[3]

Unlike the situation in the more developed countries, much of the process of modernisation and technological change which takes place in Latin America originated in more advanced countries. At the stage of purchasing technological designs, there are few differences among Latin American countries. It is at the next stage, the adaptation and improvement of imported technology to meet local circumstances, that the larger countries have considerable advantages over others. First, Argentina, Brazil and Mexico have a substantial number of highly-trained professional and technical personnel to produce an internal flow of technological knowledge in addition to the technology that is imported. Secondly, the availability of such skilled personnel facilitates the 'adaptation' of imported technology to local conditions. The Japanese mode of industrialisation has demonstrated that modern industrial success is not so much rooted in the generation of totally new technologies as in the constant adaptation and improvement of imported technological designs — first to local conditions and later for export markets. Quoting Katz, 'the industrial engineer is accustomed to the idea that every engineering design evolves and improves in time as a result of the particular circumstances in which it is used'.[4] There are many factors which may render a particular technological design inadequate in a given environment and make adjustments necessary. Some of these are of an engineering nature and concern the fact that processes have to function in different physical evironments, or with a contrasting quality of inputs. Other signs of inadequacy are more associated with economic variables; these have much to do with such factors as level of demand, conditions of supply and the most appropriate form of economies of scale. However, for whatever reason the imported technology is inadequate, it is evident that this generates a certain demand for new technological knowledge in the area of adaptation.

Although industrial technological expertise is concentrated in Argentina, Brazil and Mexico, these three countries have developed contrasting policies towards technology. Argentina has followed what could be termed a pre-Japanese model of stressing the autonomous development of technology and skilled personnel and in reducing the impact of imported technology. Perhaps as a consequence, Argentina has been able to develop an autonomous technological sector closely linked to the government and military but not able to promote a dynamic private sector, partly because of restrictions placed on imported technology. The Mexican experience reflects a policy of concentrating technological investment in crucial areas or sectors in which Mexico has a comparative advantage (oil, petrochemicals, iron and steel, newsprint) and allowing the easy import of foreign technology in other sectors; this latter part is demonstrated in the critical area of computers where no major Mexican firm has developed and where computers are either imported directly or simply assembled with minimal local content.[5]

Meanwhile, Brazil has followed a policy with definite comparisons to the Japanese model of developing technology. On the one hand, the import of new technological designs is welcome but the domestic adaptation and improvement of these designs are vigorously promoted. The result of such a policy is reflected in the development of the national computer industry where a government body, the Secretaria Especial de Informática (SEI), has supervised its growth.[6] The SEI has divided the industry into mainframe and smaller computers. Its policy towards mainframe computers has been to leave it to the multinationals, notably IBM and Burroughs, as long as they increase their manufacturing capacity in Brazil. The Brazilian government also insists that the multinationals export a substantial propotion of their output. Burroughs is due to export two thirds of the 1,200 computers it expects to build in Brazil by 1986 and IBM do Brasil has to export three out of every five computers it produces.[7] In return, Brazil strictly excludes imports of competitive machines from rival multinationals; Sperry-Univac, for example, is completely excluded from the Brazilian market.

The SEI's policy towards the manufacture of smaller computers is radically different.[8] The government reserved this growth area almost exclusively for Brazilian manufacturers as far back as 1978. Four companies from the private sector were allowed to join *Cobra Computadores e Sistemas Brasileiros* (a government-controlled company which began making computers for the navy in 1974). They were: *Edisa Eletrônica Digital*, part of the Iochpe financial group, based in Rio Grande do Sul; *Lãbo Eletrônica*, part of the Forsa group; *Sid Sistemas de Informaçao Distribuída*, a joint venture between Bradesco and the Brazilian Sharp industrial group; and finally, *Sisco Sistemas e Computadores S.A.*, part of the Maksoud engineering group. The SEI allowed these companies to begin by buying technology abroad but approval of new projects depends on their use of Brazilian expertise and locally-produced components. In this way, the SEI stresses the need for these companies to adapt and improve on imported technological designs, and to integrate their products into the Brazilian market and supply network. As a result, the five Brazilian companies have extended their range from minicomputers to microcomputers, and have been joined in this latter field by a number of new companies. By 1982, about 25 companies had approval for their projects, and as many as 15 had already begun to produce microcomputers, while many other companies have been created to produce printers, monitor screens, keyboards, disc-drives and other peripherals. According to one source,[9] the SEI is pleased with the progress made since 1978, estimating that 50 per cent of computer sales went to Brazilian manufacturers in 1982. One of the best-selling computers is the Cobra 530, with 512 K bytes of memory, completely designed in Brazil and largely made from locally produced components. Computer users in Brazil tend to be critical of the approach due to the higher costs and slightly lower specifications of Brazilian products. But the Brazilian policy towards the import, adaptation and improvement of technology in general, and computer technology in particular, was summed up by Henrique Costabile, undersecretary of the SEI in 1981: 'It is more important to have products that are made in Brazil, using technology that Brazilians dominate, than to import the last word in computers'.[10]

Thus, the three largest countries of Latin America have contrasting policies towards the domestic generation of new technology and how it complements the flow of imported technology. The most successful combination would appear to have been achieved by Brazil. It could be argued that Argentina has underestimated the benefits of a more open policy towards the import of new technologies at certain stages in its development. Meanwhile, Mexico has in more recent times preferred a freer attitude towards imported technology and has not, at least in certain sectors, actively promoted a policy of coercing local entrepreneurs to adapt and improve imported designs. In some strategic sectors, particularly in those organised by state firms such as oil refining, petrochemicals and iron and steel, there is a greater stress on the role of domestic research and development. Nevertheless, significant investment in research and development is common in all the larger firms of Brazil, Mexico and Argentina. Katz argues from an analysis of aggregate data of the three countries that 'the 100–200 major industrial concerns spend on research and development, on average, around US $150,000 a year per firm, a figure which would certainly enable them to employ an experimental-research and technological-development team which, again on average, could represent as many as six or eight professionals'.[11]

Such investment in research and development has a significant impact on the economic performance of the three countries. First of all, it has favourably affected manufacturing productivity. A number of IDB/ECLA research projects have demonstrated the great importance of minor or adaptive technological change as a source of substantial increases in manufacturing productivity. Secondly, research and development investment has begun to have a favourable impact on exports. The technological effort involved in adapting and improving upon an imported design in order to make it more suitable for local circumstances gives rise essentially to the emergence of a new product or process. The new product so developed may be more applicable for sale in Third World countries with similar local circumstances than the original imported design. One example of this is Volkswagen do Brasil's 'Brasilia' car. This was adapted from an early prototype of the Volkswagen Golf but many of the high specifications of the European model were eliminated and more sturdy and robust designs were added. The Volkswagen Brasilia, a basic sturdy model designed for Third World roads has subsequently been sold extensively to other Latin American and Third World countries — much more so than the more refined, less robust and more expensive Golf.

The sale of a vehicle adapted and improved in Brazil to neighbouring countries is an example of a new type of technological relationship. With the development of new technologies and the adaptation of imported techniques being spatially concentrated in Brazil, Mexico and Argentina, it can be predicted that those three countries will play an important future role as suppliers of technologically-sophisticated manufactures and technology itself through licencing arrangements and direct investment. Furthermore, the technological gap between the three larger countries and the rest of Latin America is likely to widen as the latest innovations and advances are rapidly transmitted to São Paulo and Mexico City where they are adapted for more general Latin American use. Indeed it could be argued that there is also a widening gap in technological expertise between Brazilian and Argentine industry.

One illustration of the widening technological gap is the contrast between the Chilean and Brazilian motor vehicle industry. The Chilean case is interesting because historically it has been able to boast the best technological expertise after the big three countries. By 1970 it had also developed a motor vehicle industry in which only 40 per cent of the value of each car produced was imported. Some significant parts firms had thus grown up. Considerable domestic technological expertise had developed in both these component firms and in the subsidiaries of international vehicle companies, notably Citroën, and vehicle models were being assembled that had been adapted to the specific Chilean conditions.[12] However, the industry had one critical failing; it was high cost. With the radical change in industrial policy of the Junta in 1974 and the adoption of a policy aimed at reducing tariffs and increasing foreign competition the technological expertise built up over a decade disappeared. The imported value of a car rose from 40 per cent to over 70 per cent, the parts industry was reduced to one third of its original size, and the research and development departments of the vehicle companies were terminated. Meanwhile, the 1970s have seen the international vehicle companies in Brazil substantially building up their research and development departments, adapting imported technologies to local conditions and exporting the 'new' models to other Latin American countries, including Chile. There has also been investment in process technology adapted to the labour-intensive nature of Brazilian assembly. The widening technological gap between Brazil and Chile in this vital sector is most acutely reflected by the argument that if Chile wished to rebuild its vehicle sector, it would be advantageous for it to purchase both the process and product technology from one of the Brazilian subsidiaries of the international vehicle companies.

There have then been radical changes in both the structure and technology of Latin American industry in the last thirty years and substantial changes in the relationship of sector and technology between Latin America and the developed world. Latin America now has the ability to manufacture in all major sectors — consumer non-durable, consumer durable, intermediate and capital goods. It also has the capacity to adapt and improve on imported technologies. The major problem of these abilities and capacities is that they are spatially concentrated within the three major countries and most notably in Brazil. Other countries, such as Chile, may be able to specialise in certain industrial sectors and develop domestic technology to improve productivity and export performance in those sectors. But the ability to produce in all major sectors and develop a wide range of home-grown technology, the necessary base for a more developed industrial society, is increasingly more spatially restricted on the Latin American continent.

References

1. See Chapter Five, for a discussion of economic integration policies
2. J. Katz, Technological Change and Development in Latin America, in R. Ffrench-Davis and E. Tironi (eds.), 1982, *Latin America and the N.I.E.O.*
3. Ibid.
4. Ibid.
5. *Latin America Weekly Report*, November 20th, 1981
6. *Latin America Weekly Report*, October 29th, 1982
7. Ibid.
8. *Latin America Weekly Report*, June 24th, 1983
9. *Latin America Weekly Report*, November 20th, 1981
10. Ibid.
11. J. Katz, Technological Change and Development in Latin America
12. R. N. Gwynne, 1978, Industrial Development in the Periphery: the motor vehicle industry in Chile, *Bulletin of the Society for Latin American Studies*, 29, p. 47

4 THE INSTITUTIONAL FRAMEWORK

Post-war industrial expansion has brought changes not only in industrial structure and technology but also in the productive agents of industrialisation. The process of industrial expansion in Latin America has been engineered through a distinctive industrial structure. This institutional structure has been referred to as the triple alliance — between state firms, national private enterprises and multinational corporations. The balance between these three institutional categories varies from country to country. Furthermore, the balance between them in any country is continually changing. In the larger countries, it would appear that national private enterprises have been losing some ground to both public enterprises and multinational corporations (in terms of their contribution to the industrial product). In the smaller countries, a reverse trend seems to be occurring with national private enterprises gaining ground at the expense of state firms and multinational corporations. Alongside these changes in balance between the members of the triple alliance, there has also been a greater differentiation of roles between state, private national and multinational enterprises. This differentiation of roles has been most apparent in the latter stages of import substitution industrialisation (ISI). As many Latin American countries have not yet reached these latter stages of ISI, such differentiation is most evident in the larger, more industrialised countries.

State Firms

It will be remembered from Chapter Two that the policy of ISI envisaged in its latter stages both a broadening of the range of local manufacturing (to include consumer durables such as the motor vehicle) and a backward integration of production (to build up the intermediate and capital goods sectors). Within this framework, it has been the role of the state enterprise to invest in such intermediate and capital goods and basic infrastructure as continued industrial expansion would require. The state adopted this role because of inadequately developed domestic capital markets which meant that only the state could provide the necessary capital for such large investments. Furthermore, as much of the investment was of a strategic nature (power, communications, steel), governments wished to exclude the participation of the multinationals.

The role of the Brazilian state in that country's rapid industrial growth since 1964 is illuminating. A large proportion of Brazil's federal investment funds are managed by the National Development Bank (BNDE). In the late 1960s, this bank was responsible for over one fifth of national capital formation. Investments were directed into steel (now 60 per cent owned by the government), oil refining (100 per cent government owned), petrochemicals, power and communications. In 1963, 65 per cent of electricity production was under private ownership; by 1982 the state had virtually complete owership of this vital sector. Powerful state companies have been created. Petrobras controls oil exploration, production and refining as well as the majority of national petrochemical installations. Siderbras was established in 1974 as a holding company to control and coordinate the Brazilian government's majority interests in eight steel companies.[1] In providing basic infrastructure and by investing in necessary though not immediately profitable industries, the Brazilian government and its state firms have undoubtedly facilitated industrial growth in many other manufacturing sectors.

Such patterns of state investment and control are not exclusive to Brazil. In 1978, the steel produced by state iron and steel enterprises made up 80 per cent of total production in Venezuela, 69 per cent in Argentina and 60 per cent in Mexico.[2] The share of state enterprises in oil refining also stands out. In Mexico, Colombia, Uruguay and Bolivia, 100 per cent of oil refining is carried out by state firms. The contribution of such firms to the processing of chemical and petrochemical products is also substantial. In Argentina and Mexico, there are large state enterprises that process basic products in these branches, while in Andean Pact countries, such as Colombia, Peru and Venezuela, all the enterprises that process basic petrochemical products are state-owned.

Increased state participation in basic industries can also be attributed to the conviction that it is a means of increasing national decision-making power in the sector. This is particularly appropriate in those countries whose governments see multinationals as posing a threat to national sovereignty and who desire complete control over production, income and capital flows in critical sectors. Nationalisation of foreign interests has been a popular policy in Venezuela in the 1970s[3] with iron ore and oil production being nationalised in 1975 and 1976 respectively. The state firm PETROVEN controls oil exploration, production and refining through four subsidiaries that took over the organisation of the four oil companies that previously controlled the industry. The state firm PEQUIVEN controls the production of basic petrochemical feedstock (ethylene, propylene, ammonia, urea) and sells these feedstocks to mainly private national firms for further elaboration.

One interesting case of increasing state involvement in industry is provided by the Venezuelan aluminium industry. The industry has become majority-owned by the government due to what could be described as a government monopoly on increased investment in the sector. Government influence is exerted through the regional development body, the *Corporación Venezolana de Guayana* (CVG), mainly because aluminium production is concentrated in the region of Guayana. In 1962, the CVG made a 50 per cent partnership with one of the six aluminium multinationals, Reynolds, to form a joint company, ALCASA, and produce aluminium at Ciudad Guayana using Caribbean bauxite and power from the CVG dam at Guri. In the early 1970s, there were disagreements between the two partners over provision of new investment and the participation of Reynolds in providing new technology. Although the partnership continued, the CVG decided that its new plant would be fully state-owned and created an aluminium subsidiary, VENALUM, for this purpose. In 1979, the VENALUM smelter at Ciudad Guayana became the largest smelter in the world with a capacity of 280,000 tons per annum; the CVG was able to guarantee high capacity utilisation after signing a contract with a Japanese syndicate to sell 140,000 tons of aluminium a year. With ALCASA's 120,000-ton capacity plant, the CVG now had the opportunity of investing in an alumina plant; approximately one million tons of alumina are required for 400,000 tons of aluminium. In 1983, a new CVG subsidiary, INTERALUMINA, completed a billion dollar alumina plant, again wholly-owned by the state. In order to feed a one million ton alumina plant, two million tons of bauxite are required; the CVG discovered a huge 500 million tons deposit of high-grade bauxite at Los Pijiguaos and created a new subsidiary, BAUXIVEN to exploit it. When the

BAUXIVEN operations are fully working at Los Pijiguaos, Venezuela will have the first fully integrated aluminium industry in the Third World with all three stages (extraction of bauxite, production of alumina, then aluminium) present in or near the Ciudad Guayana industrial complex. The six aluminium multinationals that control much of the world market in aluminium have not as yet developed a fully integrated industry in a country outside of the advanced economies. The fear of easy nationalisation has been their principal reason.

State firms have thus developed in two very distinct sectors of Latin American industrial economies. On the one hand, they have been established in the extractive industries and in the further processing of the minerals concerned. State firms in this case have developed for strategic reasons and due to a national wish for greater control over the crucial resources of a country — resources crucial for exports, taxes, the state budget, employment and the exchange rate for example. On the other hand, state firms have developed in the basic heavy industrial sectors of the more advanced Latin American economies — producing oil and steel, petrochemical feed-stocks, aluminium and electricity for a wide range of manufacturing consumers.

National Private Enterprises

In contrast to the state firm and multinational enterprise, the national private firm is characterised by great diversity in terms of size, technological level and forms of organisation. In most large and medium-sized countries, large national conglomerates have developed with a wide variety of manufacturing interests and often important tertiary functions in such areas as banking, insurance, finance, tourism, commerce and the media. Table 4.1 illustrates the wide range of interests of the six major conglomerates that existed in Chile in 1978; even the conglomerate that specialised most in manufacturing, the Grupo Angelini, had nearly 50 per cent of its subsidiary companies involved in other functions. At the other end of the size range, large numbers of small enterprises are at work filling the demand gaps left by or providing low-cost competition for the large state, national and international companies. Due to labour-intensive methods and low capital inputs, these enterprises generate a much higher proportion of employment than their

Table 4.1: Company Interests of Chile's Six Leading Conglomerates, 1978

Conglomerate	Financial Institutions	Agricultural and Timber Companies	Mining and Construction Companies	Manufacturing Companies	Commercial and other Companies	Total
Grupo Cruzat-Larrain	32	12	11	34	20	109
Grupo Javier Vial	27	6	5	15	13	66
Grupo Matte	25	7	1	7	6	46
Grupo Angelini	4	4	0	11	2	21
Grupo Edwards	23	0	0	6	6	35
Grupo Luksic	3	6	3	6	13	31
Total	114	35	20	79	60	308

Source: F. Dahse, *El mapa de la extrema riqueza* (Editorial Aconcagua, Santiago, 1979)

Table 4.2: Percentages of Assets of the Largest 300 Manufacturing Firms in Brazil and Mexico held by US, Other Foreign and National Companies, 1972

Industry	US Companies		Companies from Other Foreign Countries		National Share	
	Brazil	Mexico	Brazil	Mexico	Brazil	Mexico
Food	2%	20%	30%	6%	68%	74%
Textiles	6%	—	38%	5%	56%	95%
Metal Fabrication	4%	48%	21%	8%	75%	44%
Chemicals	34%	54%	35%	14%	31%	32%
Rubber	100%	100%	—	—	—	—
Non-Electrical Machinery	34%	36%	40%	58%	26%	5%
Electrical Machinery	22%	35%	56%	25%	22%	40%
Transportation Equipment	37%	70%	47%	9%	16%	21%
Total	16%	36%	34%	16%	50%	48%

Source: G. Gereffi and P. Evans, Transnational Corporations, Dependent Development and State Policy in the Semiperiphery: A Comparison of Brazil and Mexico, *Latin American Research Review*, 1981, 16, 3, 31–64

production levels would indicate. According to Tyler, in 1970, the 135,000 small firms in Brazil, while accounting for only 21 per cent of the total value of production, employed as much as 44 per cent of the total manufacturing workforce.[4] The organisation of most of these small firms is simple, based around a single entrepreneur or family, and many are highly susceptible to changes in the macro-economy or the policy shifts of the large companies. As a result, they have to be very adaptable and flexible in order to survive.

In terms of their relationship with multinational companies, it is often pointed out that the latter predominate in the technologically more dynamic sectors, leaving the national private firms to specialise in the more traditional industries. The 2,700 largest national private firms in Brazil, for example, account for 75 per cent of production in the non-durable consumer good sector but only 33 per cent and 45 per cent in the intermediate and metal goods machinery sectors. However, such a distinction can be misleading. National private firms are active in some very dynamic sectors in Latin America. For example, in the motor vehicle industry, national private firms have specialised in components and parts production while leaving assembly and engine production to the multinationals. In Brazil, national firms figure predominantly in the plastics, paper, machinery and microcomputer sectors. In Venezuela, the metal fabrication industry is dominated by such national firms as SIVENSA. Nevertheless, throughout Latin America, the majority of firms producing in the food, beverages, textiles, footwear, clothing, leather, cement, furniture and ceramic sectors are of national origin (see Table 4.2).

The Multinational Enterprise

The third agent of industrialisation in Latin America, the multinational enterprise, generally has its origins outside of the continent. The multinational enterprise has historically been of West European and US origin but Japanese interests have been increasing recently. The role of the multinational enterprise as an agent of Latin American industrialisation is a crucial one, not least because no other continent has received a major contribution from foreign companies in its process of industrialisation. The United States, Japan and the countries of Europe have largely industrialised through national agents, whether state or private. Other Third World countries are presently industrialising with the assistance of multinational enterprise, but Latin American countries have already developed complex and diverse industrial structures from the contributions of foreign investment. Arguably, the only close comparison would be the industrialisation of the city states of South East Asia (Hong Kong, Singapore, Taipei) where multinational enterprise has contributed to the creation of complex industrial systems.

The role of foreign companies in contributing to the industrialisation of Latin America and other less developed countries has been the focus of much theoretical work. One conceptualisation has followed Lenin's interpretation of multinational enterprise as an agent of rival imperial powers competing for control over peripheral countries. The growth of the multinational enterprise reflects the need 'to control raw-material sources and markets in order to protect their dominant position and to secure their investment even on a relatively longer-run profit perspective'.[5] Such a framework assumes an intimate link between the interests of the core country and the foreign corporation that has its origins there.

A more positive approach towards multinational enterprise and its contribution to industrial development is presented by the Nobel prize winner W. A. Lewis in his book *The Theory of Economic Growth*. Lewis sees foreign investment as providing foreign exchange, raising domestic income and increasing the labour and managerial skills of the host country.

> Domestic income increases because the undertaking pays wages and salaries to local people, buys local supplies, and pays local taxes; and these payments not only increase consumption, thereby stimulating local production, but also make it possible to have larger local savings, and also to spend on schools, medical services, and other permanent improvements. If the choice is between local capital and foreign capital the advantage may be with the former, but if, as is more often the case, the choice lies between foreign capital or leaving resources undeveloped, then there is little doubt that foreign investment plays a most useful role in providing income to pay for higher standards of consumption, of education, and of domestic improvement.[6]

The Latin American reality is somewhere between these two contrasting theoretical constructs. In the early twentieth century, there was evidence to justify the more negative aspects of international investment. The multinational companies of this era were generally involved in the extraction of raw materials, particularly petroleum (Mexico and Venezuela) and minerals (Chile and Peru). Questionable activities of these companies, such as the refusal to pay increased oil taxes after the Mexican revolution,[7] were backed up by their respective home governments who stressed the need for access to cheap sources of raw materials.

However, since the Second World War and the nationalisation of international mining interests in Mexico, Venezuela, Chile and Peru, the flavour and structure of multinational enterprise has radically changed. First of all, there has been a major increase in manufacturing as opposed to mining investment by multinationals in Latin America. In 1950, direct foreign investment (DFI) in Latin American mining and oil totalled US $2,381 million, equivalent to 53.5 per cent of total foreign investment in Latin America. By 1978, although DFI in mining and petroleum had increased to

Table 4.3: Changes in the Investment Patterns of Multinational Enterprise in Brazil, Mexico and Other Latin American Countries

	1950				1978			
	Brazil	Mexico	Other Latin American Countries	Total	Brazil	Mexico	Other Latin American Countries	Total
Total Investment (millions US $)	654	415	3,379	4,445	7,170	3,712	12,595	23,477
% of Total Investment	14.7	9.3	76.0	100.0	30.5	15.8	58.7	100.0
Extractive (%)	1.1	29.9	30.1	25.8	3.7	2.6	10.3	7.1
Petroleum (%)	17.2	3.1	32.8	27.7	5.9	1.1	25.4	15.6
Manufacturing (%)	43.6	32.0	10.7	17.6	65.3	74.1	27.1	46.2
Public Utilities (%)	21.2	25.8	20.2	20.9	0.4	0.6	2.1	1.3
Other (including Finance and Trade) (%)	16.9	9.2	6.2	8.0	24.7	21.6	35.1	29.8
Total	100.0	100.0	100.0	100.0	100.0	100.0	100.0	100.0

Source: G. Gereffi and P. Evans, Transnational Corporations, Dependent Development and State Policy in the Semiperiphery: a comparison of Brazil and Mexico, *Latin American Research Review*, 1981, 16, 3, 31–64

Table 4.4: Distribution of Foreign Direct Investment in Brazilian Manufacturing by Sector, 1979

Sector	%	Sector	%
Food	3.4	Chemicals	18.7
Textiles, Clothing	4.1	Vehicle and Auto Parts	15.3
Drinks	0.6	Machine Tools	13.6
Cellulose and Paper	3.1	Electrical and Communication	11.5
Steel	3.5	Metal Processing	7.7
Rubber	1.5	Medical and Pharmaceutical	5.2
		Others	11.8

Source: S. M. Cunningham, Multinational Enterprises in Brazil: Locational Patterns and Implications for Regional Development, *Professional Geographer*, 1981, Vol. 33, No. 1, p. 48

US $5,325 million, it only accounted for 22.7 per cent of total investment. Furthermore, mining and petroleum DFI in Mexico and Brazil accounted for only 15.6 per cent of the total.[8] The great majority of extractive industry investment was in the smaller countries of Latin America.

While mining and petroleum DFI decreased in relative importance, DFI in manufacturing increased from US $0.8 billion in 1950 to US $10.9 billion in 1978 — when it accounted for 46.2 per cent of total investment. However, such foreign investment was particularly significant in those countries that had reached the latter stages of import substitution industrialisation — notably Mexico and Brazil. Thus 68.5 per cent of DFI in Latin American manufacturing was concentrated in Brazil and Mexico in 1978.[9] As Table 4.3 shows, 74.1 per cent and 65.3 per cent of total DFI in Mexico and Brazil corresponded to the manufacturing sector in 1978. In other Latin American countries, an average of only 27 per cent of DFI was directed into manufacturing.

The majority of DFI is of US origin — 50 per cent in the large countries, 77 per cent in the medium-sized countries and 69 per cent in the smaller countries in 1976.[10] However, both European and Japanese investments are increasing. Japanese investment is still most concentrated in the processing of raw materials (steel, aluminium) but European investment is shifting to high technology sectors such as motor vehicles and chemicals.

Indeed, there is increasing evidence to show that the rapid growth of multinational investment in manufacturing in Latin America (an increase of 146 per cent between 1967 and 1976) has taken place in the more dynamic and technologically innovative sectors. In the country with nearly 50 per cent of continental foreign direct investment in manufacturing, Brazil, 77 per cent of such investment was channelled into technologically dynamic sectors in 1979 — notably chemicals, vehicles, machine tools, pharmaceuticals, communications and the electrical and medical industries (see Table 4.4). This meant that in Brazil in 1979, multinational enterprise controlled 56 per cent of total assets in the transport sector, 51 per cent in the electrical and over 35 per cent in the machinery sector. These were the three sectors that recorded the highest growth rates during the 1970s in Brazil.[11]

Multinational Enterprise and the State

Conflicts of interest between multinational enterprise and the State have been frequent in Latin America. The State is basically interested in maximising benefits for its own territory. The multinational enterprise, engaged in world-wide manufacturing operations, is interested in maximising benefits for its own international organisation. In this way, it attempts 'to find ways of achieving the economic efficiencies of world-wide specialisation of production and development within the political constraints imposed by national policies'.[12] While the State can control its own enterprises and strongly influence national private firms, it can find it difficult to exert influence over multinational enterprises. Again there are major contrasts in the experiences of large and small countries in Latin America with regard to multinational enterprises. The larger countries can often exert considerable power in bargaining with multinationals as their markets are valuable to obtain and rival companies will compete against each other for tenders.[13] Smaller countries generally have substantially less bargaining power due to their small market size and the often low level of multinational interest.

However, in both large and small countries, governments are increasingly worried about the concentration of multinational corporations in those high-technology sectors that also enjoy the highest growth rates. This is mainly because governments see growth industries in their countries being dominated more and more by organisations over which they have little control. In the booming industrial economies of the larger countries, the vital technological component of industrial growth is already strongly associated with organisations that emanate from the more developed world. This leads Latin American governments to envisage their countries' future industrial growth as being highly dependent on forces and decisions external to their territory.

The concomitant concern of governments has led some to establish strict controls over the extent and nature of foreign investment. One of the more elaborate systems of control was that known as Decision 24 of the Andean Group (Venezuela, Colombia, Ecuador, Peru, Bolivia, Chile) in 1974; the most controversial aspect was a 14 per cent profit and dividend remittance ceiling on registered capital for multinational companies. The major effect of Decision 24 was to reduce foreign investment and the scope of new industrial projects in the region. This was particularly the case in Venezuela which could have anticipated a significant expansion of foreign manufacturing investment in the 1974–7 period due to its large oil revenues and its policy of investing a large proportion of them in industrial development. However, foreign investment remained low during the period. In 1977, Venezuela introduced a much less restrictive foreign investment law, Decree 2442, allowing increased profit remittances, removing the elaborate financial controls and extending the time limits on transfer of technology contracts.[14] As a result, foreign investment in Venezuelan manufacturing has expanded more rapidly in subsequent years.

Controls, therefore, tend to restrict foreign investment and, because such investment tends to be in the more dynamic manufacturing sectors, industrial growth as well. Lack of controls can also have adverse effects, permitting excessive royalties and profit remittances, the financial juggling of costs across national boundaries and an over-reliance on locally-generated capital. Lack of control can also affect the structure of industry, given that some world industries are now dominated by a small number of large corporations that can be described as forming an interacting oligopoly.[15] In this context, a locational decision taken by one corporation can generate similar responses from other members of a world oligopoly.[16] This can mean that once one member of a world oligopoly has decided to locate a plant in a new market, other corporations will follow. In Latin America, such oligopolistic reaction has been notable in the motor vehicle and tyre industries.[17] In large countries, such mass entry can lead to competition between firms utilising much of their installed capacity. But in small countries the tendency has been for a large number of firms to produce at low capacity and high costs, thus considerably adding to problems of economies of scale. Despite this, few countries have atttempted to control the number of multinational firms locating in their country — even within the framework of import substitution industrialisation.[18]

Multinationals and Export Production

Chapter Two demonstrated that the last decade has generally witnessed a shift to more open economies in Latin America and pointed out that this has had major implications for industrial performance and direction. On the one hand, the larger countries, due to their production of industrial goods being internationally competitive, have been able to expand significantly industrial exports without major increases in manufactured imports. On the other hand, the smaller countries, with few industrial sectors internationally competitive, have been unable to expand their industrial exports and have often suffered major increases in manufactured imports.

The multinational companies have responded to these changes in industrial performance and direction. It would appear that they have attempted to increase their participation in the larger countries, attracted by the large domestic markets and good export records. In the smaller countries, they have been distinctly less interested. Indeed as opening up the economies of many small countries has caused the decline (and even disappearance) of many industrial sectors, the multinationals involved in these sectors have often decided to leave or sell out to state or private national firms.

Thus the balance of the triple alliance has been changing in the last decade, but the nature of the change is closely linked to the size of country. In the larger countries, the multinational companies have been distinctly ambitious for a greater involvement in industrial development. As a result, the contribution of the multinationals to the industrial product is often increasing at a greater rate than that of private enterprises. However, in the smaller countries, where multinationals have been less enthusiastic in the last decade and where many have been pulling out, the reverse is often the case. In terms of the contribution to the industrial product, private national enterprises have been gaining ground at the expense of multinationals.

These points can be illustrated by examining two examples — the record of multinational interest in Chile since 1973 and recent developments in the Latin American motor vehicle industry.

Multinational Investment in Chile Since 1973

Since 1973, conditions in Chile would appear to have been conducive to significant investment from multinational companies. The various economic teams of General Pinochet have followed a constant line in aiming to integrate the Chilean economy more fully into the world capitalist economy, in emphasising the role of the private capitalist as the main motor of economic growth, and in welcoming foreign capital. The desire to encourage foreign investment was a major reason for Chile's withdrawal from the Andean Group in 1976, due to their strict controls on foreign investment. Chile's own foreign investment law (Decree Law 600 passed in 1976), was a much more liberal document, allowing freedom for profit remittances and technology transfer payments. There were no special company taxes over those paid by national companies. Furthermore the mid-1970s saw great opportunities for investment. The Marxist government of Salvador Allende (1970–3) had nationalised most foreign companies and intervened in or taken over most of the hundred largest Chilean companies. The Pinochet government was selling these back to private enterprise often at very low prices. However, at the same time, the economic teams of General Pinochet were consistent in opening up the economy to outside competition. According to Congdon[19] the move from autarkic to open policies in Chile 'was perhaps the most rapid and complete ever implemented'. Exports increased from $850 million in 1972 to $4,722 million in 1980. Imports increased from $1,103 million in 1972 to $5,777 million in 1980. The ratio of trade to GDP increased from 24.7 per cent in 1972 to 45.6 per cent in 1980. Non-traditional exports (principally timber, fish and fruit) grew from $71 million in 1973 (5.7 per cent of total exports) to $1,629 in 1980 (34 per cent of total exports). The development of these new activities gave impetus to the economy which registered an average annual growth rate of over 7 per cent between mid-1976 and mid-1981.

Conditions seemed favourable for a major return of the multinational enterprise to Chilean manufacturing. But such a return did not materialise. The paucity of multinational enterprise was most evident in the former import-substituting sectors. These sectors had been protected by high tariffs from foreign competition for many years. Soon after the coup, in 1973, Pinochet's First Finance Minister, Gotuzzo, began to bring tariffs down to between 25 and 35 per cent — quite high but much reduced from earlier levels of over 100 per cent. However in 1977, the Finance Minister, Sergio de Castro, announced that there would be a 10 per cent uniform tariff by June 1979 (with the basic exception of motor vehicles over 850 c.c.).

Multinational interest in the import-substituting sectors of Chilean industry has as a result been small — even in the favoured motor vehicle sector where a 10 per cent uniform tariff was not planned to come into force until 1986. In 1974, there were eight vehicle plants in Chile. Their multinational links were with Ford, Peugeot-Renault, Fiat, Nissan, British Leyland, Chrysler, General Motors and Citroën. The Pinochet government put out tenders to those companies wishing to stay in Chile. Ford, Nissan, British Leyland and Chrysler were uninterested from the outset. Citroën had been working in a joint venture with the government holding company, and did not wish to increase its capital participation. The Peugeot-Renault plant stayed under the control of local interests. Only Fiat and General Motors bought back or resumed control of their plants. With serious competition from Japanese imports (particularly in small cars of less than 850 c.c.), Citroën and Fiat terminated production in 1982. This is the best documented decline of multinational interest in Chilean industry over the last few years. Japanese imports of Datsun, Daihatsu, Suzuki, Toyota and Subaru cars accounted for 40 per cent of the Chilean market in 1982. The two remaining companies Peugeot-Renault and General Motors, have only a 25 per cent market share and plan to terminate production by 1986, unless the rules are changed.

Direct foreign investment has been attracted into Chile — but the manufacturing sector has not been prominent. Between the passing of Decree Law 600 in 1976 and end-1982 US $1,789 million had been invested in Chile by foreign companies. However, in sharp contrast to the Brazilian case, only 23 per cent of this investment had been in manufacturing, whereas 40 per cent of this investment had been in mining and 30 per cent in services. Much of the $412 million invested by foreign companies in Chilean manufacturing industry between 1976 and 1982 had been in export-oriented industry such as fish-meal and metal refining. Other major manufacturing investments have been in petrochemicals (Esso,

Dow, Shell), cement (Portland, Blue Circle), tyres (Firestone, Goodyear) and food products (Ambrosoli, Coca Cola). Only five of the largest 50 private firms in Chile in 1978 (assets greater than $20 million) were foreign-owned.[20]

When plants and firms were sold back to private enterprise in the mid-1970s it was national rather than multinational companies that bought up the manufacturing installations. For example, Unilever was not interested in re-purchasing their old Chilean subsidiary, Indus Lever, and one of the powerful economic groups in Chile, the Grupo Vial, gained control. A small collection of powerful economic groups emerged in Chile. Through access to cheap international finance and control of banks in Chile, they were able to purchase from government, uninterested multinationals and smaller firms, significant amounts of Chile's manufacturing base. By end-1978, the largest group, Cruzat-Larrain, owned assets in 34 manufacturing companies, and the Grupo Vial controlled 15 manufacturing companies. The next four largest groups had interests in a further 38 manufacturing companies. As a footnote, one could add that the two most powerful groups, Cruzat-Larrain and Vial, needed to take out huge loans on the international currency markets in order to finance these empires. When the peso was devalued against the dollar in mid-1982, their dollar debts rose against their peso assets, and they both met financial ruin in January 1983, when they were taken over by government receivers. As a consequence, the government have once again to decide how to dispose of a significant number of manufacturing companies.

The Chilean example has served to demonstrate how multinational enterprise has become less significant in the manufacturing sector of a small country 'opening-up' its economy to outside competition. Instead, another element of the 'triple alliance' has become more prominent in manufacturing. Such a pattern has also been witnessed in other small countries opening up their economies in Latin America (Uruguay and Peru). Even a country of moderate size, such as Argentina, has noted such a multinational response as it opened up its economy under President Videla and his economy minister, Martinez de Hoz. Such trends can be identified if one examines the Latin American motor vehicle industry and in particular compares multinational behaviour between Brazil and Argentina.

The Motor Vehicle Multinationals in Brazil and Agentina

One study on Argentine industry in the late 1970s concluded that:

> there is little evidence of increased foreign ownership in manufacturing industry, largely because Argentina has been seen as an uncertain and shrinking market. Foreign capital may be attracted to specific projects, but its main form of participation during the Martinez de Hoz years was through the financial system.[21]

Such a conclusion is more than borne out by the experience of the motor vehicle industry since 1976. In that year there were still eight major producers in Argentina, producing between them 186,000 vehicles (Chrysler, Citroën, Fiat, Ford, General Motors, Mercedes Benz, Peugeot and Renault).[22] Martinez de Hoz then began to submit companies to increased competition from imported vehicles, by lowering tariffs. Citroën and General Motors closed down their plants and switched to becoming importers. Chrysler sold its Argentinian operations to Volkswagen who have been studying ways of linking the plant to its large Brazilian works. Fiat began to make large losses and sold majority control to Peugeot, who in turn sold a majority stake to a local group, Francisco Marci; the new group was labelled Sevel and is now the 14th largest firm in Argentina. Ford (4th largest firm in Argentina) and Renault (6th largest) are now the only multinationals operating car plants in Argentina and Mercedes Benz (17th largest) concentrates on trucks.

The major problem of the Argentinian motor vehicle industry (as with the Chilean) has been its inability to achieve significant economies of scale within an import substitution framework. In 1965, its output was higher than that of Brazil and in 1973, nearly 300,000 vehicles were produced. Subsequently, however, production decline has been the rule with 1976 production levels of 190,000 and 1982 levels of 130,000. The near-total dependence of the Argentine motor vehicle industry on the home market has caused it to become locked into a downward spiral, where a declining market signifies lower production and consequent reductions in economies of scale, which in turn act to increase prices and lower demand still further.

In contrast, the Brazilian domestic market managed to increase to an average of over 800,000 vehicles a year by the latter half of the 1970s, a figure that allowed more than one manufacturer to benefit from significant economies. As a result, and in contrast of Argentina, multinationals have been actively competing to stay producing and enter the Brazilian market. In 1976, there were five significant vehicle producers in Brazil — Chrysler, Ford, General

Motors, Mercedes Benz (mainly trucks and buses) and Volkswagen when Fiat entered the market building a large plant in Minas Gerais. Subsequently, Chrysler sold out to Volkswagen in 1978. Five multinationals found it difficult to achieve the requisite economies of scale in the Brazilian market, particularly when the domestic market fell to below 600,000 in 1981. In October 1982, after Fiat had announced that it was pulling out of all its Latin American operations apart from Brazil, both Robert Garrity (President of Ford do Brazil) and Joseph Sanchez (President of General Motors do Brazil) stated that the Brazilian market was only large enough for four big manufacturers, not five.[23] The obvious target for the remark was Fiat who, since their entry into Brazil in late 1976, had not managed to carve out a significant market share in Brazil. In 1982, its domestic sales were only 44,000 and it survived by exporting 103,000 cars back to Europe. As a result Fiat made significant losses in Brazil in both 1981 and 1982, but according to the head of Fiat's international operations in 1983, Vittorio Ghidela, 'The key country in the South American car business will continue to be Brazil, and we are determined to stay there, even if the losses continue'.[24]

Vehicle multinationals have therefore differentiated between large and smaller countries as Latin American economies generally have become more open. During the closed import-substitution phase of industrial development, multinationals achieved greatest success in terms of reducing costs and increasing demand in the larger countries of Latin America. Their record in the smaller countries was generally much poorer. As Latin American economies became more open in the 1970s, multinationals have responded by staying and competing fiercely in the larger countries. In the smaller countries, they have either virtually left (Chile) or have left the market to a much smaller number of producers (Argentina). As a result, it is the larger countries that appear to have benefited most from the multinational contribution to the triple alliance. In the smaller countries, the multinationals that remain still basically focus on the domestic market and exports of vehicles and components are minimal. In the larger countries, however, the multinationals (as with the Fiat case) are forced to export in order to survive in the market they regard as vital. In the case of Brazil, the need for the multinationals to export meant that the full effect of the 1981 crisis (when domestic sales fell by 40 per cent) was reduced when vehicle exports rose by 37 per cent. While multinational vehicle exports had been just over 10 per cent of domestic sales in 1980, in 1981 this ratio rose to nearly 25 per cent. In the smaller countries, meanwhile, a domestic crisis in sales is normally met by the exit of one of the multinational producers.

For this reason the triple alliance is evolving differently in Latin America during *apertura* — depending on size of country. In the larger countries (most notably Brazil and Mexico), multinational interest in production is high and government policy aims to control and often restrict multinational participation — as with the mini computer industry in Brazil. Governments concentrate on preserving the interests of the other two sections of the triple alliance. In smaller countries, on the other hand, multinational interest has been lower and even governments enthusiastic about the involvement of multinational enterprise have been unable to attract much direct foreign investment in manufacturing. In these countries, the national private enterprise has asserted itself as the major element of the triple alliance — as in both Chile and Argentina.

Latin American industrialisation, in contrast to the industrialisation of Europe, United States and Japan, has developed within the institutional framework of a triple alliance of state firms, multinational corporations and national private enterprise. Within the process of industrialisation, each has developed a well-defined role. To a certain extent, the three contrasting roles have been compatible, particularly in the more mature industrial countries of Brazil and Mexico. Nevertheless, the fact that multinational enterprise is playing such a critical role in the development of the more dynamic and technologically innovative sectors may have serious implications for the long-term stability of the industrialisation process.

References

1. J. P. Dickinson, 1978, *Brazil*. (Dawson, Folkstone)

2. United Nations Economic Commission for Latin America (UNECLA), 1979, *International cooperation for industrial development in Latin America*. (Second Latin American Conference on Industrialisation, Cali, Colombia)

3. P. E. Sigmund, 1980, *Multinationals in Latin America: The Politics of Nationalisation*. (University of Wisconsin Press, Madison)

4. W. G. Tyler, 1981, *The Brazilian Industrial Economy*. (Heath, Lexington, Mass.)

5. I. Roxborough, 1979, *Theories of Underdevelopment*. (Macmillan, London)

6. W. A. Lewis, 1955, *The Theory of Economic Growth*. (George Allen and Unwin, London), p. 258

7. P. E. Sigmund, *Multinationals in Latin America*, p. 51
8. G. Gereffi and P. Evans, 1981, Transnational Corporations, Dependent Development and State Policy in the Semiperiphery: A Comparison of Brazil and Mexico, *Latin American Research Review*, 16(3), pp. 31–64
9. Ibid.
10. UNECLA, *International cooperation for industrial development*
11. S. M. Cunningham, 1981, Multinational Enterprises in Brazil: Locational Patterns and Implications for Regional Development, *Professional Geographer*, 33(1), p. 48
12. G. S. Gunderson, 1979, The Worldwide Corporation — an economic catalyst, *IGU Commission on Industrial Systems*, Rotterdam Symposium
13. R. Vernon, 1978, *Storm over the multinationals: the real issues*. (Macmillan, London)
14. P. J. West, 1978, Venezuela: foreign investment policy, *Bank of London and South America Review*, 12(3), p. 118
15. F. T. Knickerbocker, 1973, *Oligopolistic Reaction and the Multinational Enterprise*. (Harvard University Press, Cambridge, Mass.)
16. R. N. Gwynne, 1979, Oligopolistic reaction, *Area*, 11(4), pp. 315–19
17. P. J. West, 1977, *The Tyre Multinationals: A Study of Foreign Investment and Technology Transfer in Latin America* (Unpublished doctoral dissertation, University of Sussex)
18. R. N. Gwynne, 1978, Government Planning and the Location of the Motor Vehicle Industry in Chile, *Tijdschrift voor Econ. Soc. Geog.*, 69, pp. 130–40
19. T. G. Congdon, 1982, "Apertura" policies in the cone of Latin America, *The World Economy*, 5, 2, pp. 133–48
20. F. Dahse, 1979, *El mapa de la extrema riqueza*. (Editorial Aconcagua, Santiago)
21. J. Schvarzer, 1983, *Cambios en el liderazgo industrial argentino en el período de Martinez de Hoz*. (CISEA, Buenos Aires)
22. R. N. Gwynne, 1978, The Motor Vehicle Industry in Latin America, *Bank of London and South America Review*, 12, 9, pp. 462–71
23. *Latin America Weekly Report*, October 29, 1982
24. *Latin America Weekly Report*, April 2, 1983

PART TWO

SPATIAL PERSPECTIVES ON INDUSTRIALISATION

5 REGIONAL ECONOMIC INTEGRATION

When Latin American countries achieved independence in the first half of the nineteenth century, the highly centralised Spanish empire on the Latin American mainland divided into sixteen separate countries. The Portuguese Empire, which had been much more loosely controlled by the Portuguese Crown, remained intact as one country. The leader of the Spanish American independence movement, Bolivar, dreamed of a United States of Spanish America, similar to that of North America. Unfortunately, the Spanish American empire was the only mainland American empire to disintegrate around a series of provincial and urban interest groups. The consequent small size of Spanish American countries has consistently been a problem for economic and political development. In terms of industrialisation, small market size has proved a major bottleneck, as Chapters Two and Three have pointed out. This was particularly true in the period of import substitution industrialisation when the production of consumer products for small markets was seriously hampered by a lack of economies of scale and high costs.

One way out of this problem of small market size was for small countries not to see their problem individually but to group together and form, through a process of economic integration, a much larger market. By countries coming together in some form of economic cooperation, industrialists within those countries would have access to a considerably larger market than if the countries remained totally independent. There are a variety of factors that can assist the process of economic integration in promoting industrialisation in small less developed countries.

First, the degree of integration is an important variable. As a general rule, one can state that the greater the degree of integration, the greater the possibilities for industrial development. The simplest form of economic integration is the Free Trade Area, where tariffs among participating countries are abolished but each country retains its own tariff against non-members. However, for the purpose of promoting manufacturing growth in small Latin American countries, the Free Trade Area is inadequate as most industrial goods can still be imported from developed countries or cheap international sources by the various member countries. According to Griffin, another significant difficulty with this arrangement is that 'low tariff member countries are tempted to import goods from non-members for re-export to their partners'.[1] A better solution is the formation of a Customs Union. A Customs Union differs from a Free Trade Area to the extent that member countries agree to a common tariff level for third countries. High common external tariffs on manufactured goods can make their import expensive and allow industrial plants from within the Customs Union to supply the total market. Closer integration is achieved under a Common Market in which restrictions on factor movements such as labour and capital are eliminated. With greater integration, it is possible to harmonise industrial development between the various countries concerned — for example, in terms of a regional investment policy or a regional allocation of industry policy in which each member country is awarded a certain industrial sector upon which to concentrate its industrial investments. Thus, with increasing integration, there is a greater potential for industrial development due to the possibility of creating high external protective tariffs and coordinating common programmes of industrial development for individual sectors within the group of countries concerned.

Secondly, industrial development within a group of countries is enhanced if some level of manufacturing has already been achieved. In this way, firms from the different countries can compete with each other for the larger market that they can now reach. In competing, these firms will attempt to improve efficiency and productivity in order to increase their share of the new, extended market. However, if only one firm exists within the new grouping, or if no firms exist and one has to be created, the firm in question will enjoy a monopoly and there will be no onus on it to become more efficient. Complementarity, where one country exclusively produces one type of good and another country controls the exclusive manufacture of another, means that there is no pressure to become more efficient. Some degree of competition between firms of different countries is necessary to create a more dynamic and efficient industry for the market as a whole. Unfortunately, beyond certain basic manufacturing sectors, such a condition is uncommon among groupings of small countries in Latin America or other less developed regions — indeed the small countries often come together in order to stimulate wider industrial development.

Thirdly, industrial development in a grouping of countries is enhanced if they are close together and if a reasonable infrastructure interconnects them — in particular, a good road system. A final point would introduce the question of foreign capital. The attraction of foreign firms and investment to an economic grouping is enhanced by a perception of permanence. If a grouping of countries has numerous difficulties over integration and seems unlikely to stay together for long, foreign firms are less willing to locate a new plant, as, with the break-up of integration, the foreign firm may well be left serving only the domestic market of the country in which it is located.

Recent attempts at economic integration in Latin America began with the highly ambitious Latin American Free Trade Association (LAFTA) in 1960, which attempted to liberalise trade between Brazil, Agentina, Chile, Uruguay, Mexico, Peru, Paraguay, Colombia, Ecuador, Venezuela and Bolivia. Some liberalisation of trade did take place, but most benefits were concentrated in Brazil, Mexico and Argentina. More successful as a scheme of economic integration was the Central American Common Market (CACM) also created in 1960. The CACM consisted of five small countries at comparable levels of development — Guatemala, El Salvador, Nicaragua, Costa Rica and Honduras. This enlarged market gave manufacturers a market of ten million, up to ten times larger than previous national markets. Furthermore, rapid advance towards integration took place with a Customs Union virtually being formed by 1966, when intra-regional free trade existed on all but 8 per cent of items and common external tariffs had been agreed on the great majority of items.

When LAFTA had been seen to stagnate as a scheme of economic integration, five of its smaller members geographically attached to the Andes (Chile, Bolivia, Peru, Ecuador and Colombia) formed the Andean Pact in 1969. The main motivator in its formation was Chile whose industrial development by import substitution had been severely constrained by small market size. The prospect of an enlarged market of seventy million attracted all five countries and later Venezuela in 1973. Unfortunately the pace towards integration was notably slower than in the Central American Common Market. Intra-regional tariffs were still set as high as 26 per cent in 1980.[2] This means that the Andean Group had not effectively become a Free Trade Area by 1980.

However, both the Andean Pact and CACM have suffered from the basic problem of schemes of economic integration — the unequal distribution of benefits among the countries concerned. As schemes of economic integration develop, there follows a marked process of spatial concentration of benefits in the more developed and larger countries of the grouping. In terms of industrial benefits, the locational policies of multinational firms are important in this respect. Such firms tend to locate in the largest and most developed market of the grouping for two very powerful reasons. On the one hand, they will have access to the best conditions of the grouping in terms of labour supply, industrial suppliers, local capital markets and services. On the other, they will have a good 'fall-back' position if the scheme collapses or deteriorates in any way. Although no longer benefiting from an extended, international market, they would nevertheless have access to the largest market of the grouping.

As a result, industrial development will proceed faster in the larger countries of the grouping due in part to the locational decisions of both multinational and related firms, whereas the smaller and poorer countries will suffer from trade diversion. One of the theoretical advantages attributed to economic integration is that of trade creation, when a low cost supply replaces a high cost supply. However, in groupings of less developed countries, trade diversion is more often the result — that is, a low cost supply replaced by a high cost supply. A Central American example, that could be called the 'Honduras syndrome', illustrates the point well. Before the CACM was established, Honduras purchased tyres from the cheapest international source available — probably from Goodyear and Firestone plants in the south of the United States. However, after 1966, when the only tyre industry of the CACM, the Goodyear plant in Guatemala, was established, Honduras had to buy tyres from a plant that became a relatively high-cost producer of tyres. Costs were high because of low economies of scale and the monopoly position that the plant held within the CACM. Thus, Honduras had effectively suffered from a case of trade diversion with a high-cost supply replacing a low cost supply. Evidently, Honduras would have to extract some definite advantages from the process of economic integration in order to counterbalance the increasing costs suffered as the result of numerous cases of trade diversion.

Both the CACM and Andean Pact have attempted to create complex planning arrangements and harmonisation policies in order to rectify imbalances and give some advantages to the poorer countries. The 1960 General Treaty of the CACM approved a regional allocation of industry scheme by which large-scale basic industries would be allocated to individual member countries. Each industry would benefit from the maximum economies of scale that the CACM could offer as no rival firm would be permitted to compete in the 'allocated' firm's regional market for at least ten years. In order that all CACM countries should benefit, each country would receive one of these industries and no country would receive a second industry before all five countries had received one. The first plant, Guatemala's tyre and tube plant, began operating in 1966. It had taken Goodyear three years and the Guatemalan government two years of negotiation before the final ratification took place.[3] El Salvador, in particular, was against the scheme, presumably because it preferred a more spatially concentrated form of industrialisation within the CACM. In any case, Honduras did not receive an 'allocated' industry. Indeed, after the Guatemalan plant, only Nicaragua managed to build an 'allocated' industry — a caustic soda and chlorinated insecticides plant.

The failure of the regional allocation of industry scheme in the CACM was not only due to the long negotiating process (all five countries had to approve each scheme) but also due to the CACM approving in 1963 an arrangement called 'The Special System for Promotion of Productive Activities'.[4] According to the arrangement, very much developed at the initiative of El Salvador, the five countries would agree on the regional desirability of a list of industrial products, without specifying the precise location of future production. The first plant in each product sector which could prove that its effective capacity could cover 50 per cent of regional demand would be granted additional tariff protection against extra-zonal imports. The industrial growth promoted by this scheme very much favoured the two countries with the larger industrial structures, namely Guatemala and El Salvador. By 1978, they were responsible for 60 per cent of intra-regional exports.

However, the failure of the CACM's harmonisation policies to promote industrial development in Honduras and thereby to counterbalance that country's increasing disadvantages of membership, led to the withdrawal of Honduras from full membership of the Market at the end of 1970. The withdrawal followed the outbreak of war between Honduras and El Salvador, the neighbour that Honduras regarded as being the major beneficiary of the Market. Subsequently, the CACM has stopped being a full Customs Union with the regional programmes of industrial development being abandoned and a series of bilateral trade agreements being signed between member countries.

The problems of the spatial concentration of benefits in the more powerful countries and the need to protect the interests of the poorer countries were also recognised by the Andean Group's 1969 Charter, the Cartagena Agreement. Bolivia and Ecuador were granted special concessions in terms of both intra-regional trade and the common external tariff.[5] Furthermore, priority was to be given to Bolivia and Ecuador in the allocation of industries under the regional sectoral programmes. However, in order to attempt to equalise the benefits of regional industrial programmes among all countries, highly bureaucratic negotiating machinery and complex regulations were developed. Such complexity surrounds the implementation of regional sectoral programmes that little real advance has been made in the creation of regionally-integrated as opposed to nationally-integrated industries. The implementation of the regional motor vehicle programme provides a good example of these problems.[6]

In accepting the Automobile Programme (Decision 120) in Quito in September 1977, the Andean Group gave special concessions to both Bolivia and Ecuador. Decision 120 stated that both Bolivia and Ecuador should be spared from applying the Common External Tariff on both vehicles and components until the end of 1988; the other three countries should comply by the end of 1983. In terms of intra-regional trade, Bolivia and Ecuador can put extra tariffs on cars imported from other Andean Group countries until the end of 1988. The other three countries, however, have to import any regionally-approved vehicle from Bolivia or Ecuador without tariffs from the beginning of 1982. Theoretically, therefore, Ecuador and Bolivia have considerable cost advantages in the production of vehicles for the Andean market between 1982 and 1988. They can import all parts except those of the basic model (engine, axles, gearbox and steering system) free of tariffs and then produce vehicles that can enter the whole Andean market without extra tariffs being placed on them.

Although Bolivia and Ecuador do enjoy substantial advantages in the regional automobile industry in the first six years, in order to enjoy these advantages, they need to possess an established industry in a position to expand. Unfortunately Bolivia had still not decided in 1982 on which international vehicle corporations would be awarded their small pick-up and heavy truck allocations (see Table 5.1). Bolivia's political difficulties with the Andean Pact countries and the increasing trade and technology links with Brazil are causing the country to distance itself from the Pact and delay many important decisions.

Ecuador, despite having no vehicle industry in 1981, has awarded its small car allocation to Volkswagen and its pick-up to General Motors. Construction of plants will begin when all Andean Pact allocations have contracts and have been signed by all parties. However, such signing is now being delayed by two of the more powerful countries, Peru and Venezuela, who can use the highly bureaucratic negotiating machinery to delay the signing of contracts of other countries' allocations as well as their own. Some of the complex arrangements of the Andean Group Automobile Programme can be seen from Table 5.1. Apart from the relatively straightforward allocation of basic models to countries, the pattern is complicated by the multiplicity of assembly and co-production agreements that exist. Assembly agreements permit a country that has received the allocation of a vehicle category, and has selected its basic vehicle, to authorise another country to assemble that vehicle. Meanwhile, co-production agreements permit member-countries to specialise in the production of certain parts and to interchange them. For example, Venezuela has a co-production agreement with Ecuador for the category of B1.2 pick-up trucks; GM, who won the contract, will produce engines in Venezuela and gearboxes in Ecuador, but will assemble the pick-up in both countries.

The limited progress of integration in the Andean Group since 1969 has been associated with a relatively even distribution of those benefits that have accrued. For example, between 1971 and 1981, the value of regional exports (at constant 1971 prices) more than doubled in Ecuador and Venezuela and more than trebled in Colombia and Peru; only Bolivia stood apart with an effective 20 per cent decline in the value of regional exports for the period.[7] However, the complex machinery that has been created to equalise benefits and prevent inequalities has also acted as a barrier to the complete liberalisation of trade within the Group, the adoption of a common external tariff and the execution of the industrial development programmes. Not only the motor vehicle but also the metal-

Table 5.1: Allocations, Assembly and Co-Production Agreements within the Andean Pact Automobile Programme, 1983

Category		Bolivia	Colombia	Ecuador	Peru	Venezuela
A1	(Car, up to 1,050 cc)	—	Allocation (Renault)	—	—	—
A2	(Car, (1,050–1,500 cc)	—	Allocation (Renault)	Allocation (Volkswagen)	—	—
A3	(Car, 1,500–2,000 cc)	Assembly Agreement with Venezuela	Assembly Agreement with Venezuela	—	Allocation of 2 engines (Volkswagen)	—
A4	(Car, over 2,000 cc)	Assembly Agreement with Venezuela	Assembly Agreement with Venezuela	—	—	Allocation (General Motors)
B1.1	(Pick-up, (up to 3 tonnes)	Allocation	Assembly Agreement with Bolivia	—	—	Assembly Agreement with Bolivia
B1.2	(Pick-up, 3–4.6 tonnes)	—	—	Allocation (General Motors)	—	Co-production Agreement with Ecuador
B2.1	(Truck, 4.6–6.2 tonnes)	Assembly Agreement with Peru	Assembly Agreement with Peru	—	Allocation (Volkswagen)	—
B2.2	(Truck, 6.2–9.3 tonnes)	Allocation Agreement with Peru	Assembly Agreement with Peru	—	Allocation (Volkswagen)	—

Table 5.1 — *continued*

Category		Bolivia	Colombia	Ecuador	Peru	Venezuela
B3	(Truck, 9.3–17 tonnes)	Allocation	Allocation	Assembly Agreement with Venezuela	Allocation	Allocation
B4	(Truck, over 17 tonnes)	—	Assembly Agreement with Venezuela	Assembly Agreement with Venezuela	Allocation (Volvo)	Allocation
C	(Jeep, up to 2.7 tonnes)	Assembly Agreement with Colombia	Allocation (Renault)	Assembly Agreement with Venezuela	Co-production Agreement with Colombia	Allocation

Note: Not all companies inserted above have signed their final Andean Group Contracts and therefore their allocations are not always definite

working, petrochemicals and steel programmes are suffering from disagreements between the five countries about the final nature of each programme. The more industrialised countries (notably Venezuela) want high tariffs to protect their industries, while the less industrialised (notably Ecuador and Peru) demand low tariffs.

From such an analysis, one must be sceptical about the possibilities for industrialisation in small countries within the framework of schemes of economic integration. There seem to be two rather stark alternatives. On the one hand, one has a position in which checks and balances of the integration scheme are reduced, leading to the spatial concentration of benefits and the demise of the scheme due to the exit of the country (or countries) which has or have not shared in the benefits. Alternatively, a more elaborate system of checks and balances permits extended procrastination of the execution of industrial programmes, and the eventual stagnation of programmes as countries gradually reorientate their industrial policies back to the national framework.

The sixteen countries of the Spanish American mainland that divided politically after independence have been unsuccessful in their attempts to integrate for mutual economic benefits. Economic development is very much conducted within the confines of national boundaries. International links and communications of all types are still poorly developed. Road links are particularly poor. There is no major road linking Pacific and Atlantic at the latitude of Latin America's most dynamic region, SE Brazil. There is still not a continuous road link from Central to South America. Without the completion of such infrastructure of integration, the economic, political and geographical 'disintegration' of Latin America, inherited from the nineteenth century, will continue into the twenty-first.

References

1. K. Griffin, 1969, *Underdevelopment in Spanish America.* (George Allen & Unwin, London), Chapter Six

2. D. Hojman, 1981, The Andean Pact: Failure of a Model of Economic Integration? *Journal of Common Market Studies*, 20(2), pp. 139–60

3. P. C. Schmitter, 1971, Central American Integration: Spill-over, Spill-around or Encapsulation? *Journal of Common Market Studies*, 9(1)

4. Ibid.

5. D. Morawetz, 1974, *The Andean Group: A Case Study in Economic Integration among Developing Countries.* (MIT Press, Cambridge, Mass.)

6. R. N. Gwynne, 1980, The Andean Group Automobile Programme: an interim assessment, *Bank of London and South America Review*, 14(3), pp. 160–70
7. D. Hojman, The Andean Pact

6 THE SPATIAL CENTRALISATION OF INDUSTRY

The spatial emphasis up to this point has been concerned with how industrialisation differs among Latin American countries. In this chapter, however, the analysis will concentrate on the spatial development of manufacturing industry *within* individual Latin American countries and, in particular, on the dominant spatial process of industrialisation in Latin American countries — the centralisation of industry.

The centralisation of industry in the major city of a country is a well-established empirical fact in Latin America. In the two most advanced industrial nations of Brazil and Mexico, over 50 per cent of their respective manufacturing production and employment is located in the São Paulo[1] and the Mexico City[2] agglomerations. In the less advanced industrial nations, spatial concentration of industry in the major city is even greater. Two thirds of Argentina's industrial employment is located in the Buenos Aires metropolitan area[3] and 75 per cent of Venezuela's manufacturing workers are concentrated in the major axial belt of Valencia — Caracas — Guatire.[4] Meanwhile, 70 per cent of Peruvian[5] and 57 per cent of Chilean[6] manufacturing employment is centred in Lima/Callao and Santiago respectively. In the smaller, less-industrialised countries where only the first stage of import substitution industrialisation has been completed, spatial concentration of industry in the primate city is greater still.

How can such accentuated concentration of industry in central cities be understood and how can the strength of the centralising process in manufacturing be gauged? The answers to such questions will be relevant to the problem of decentralisation which will be discussed in the following chapter. One way of attempting to comprehend and gauge the processes behind the centralisation of industry is to analyse the phenomenon with reference to the body of location theory that has been developed, mainly in countries of Western Europe and the United States, in the twentieth century.

Location Theory

Such an approach of applying location theory to the centralisation of industry in Latin America does have certain difficulties. First, the majority of location theories have been developed in Western Europe and the USA and if these theories are not as Weber[7] termed 'pure' (i.e. applicable to any political or economic system), it is possible that their relevance may be limited in Latin America. A further problem concerns the method of applying location theory to the problem. Can location theory be integrated into a theoretical whole as Lloyd and Dicken maintain?[8] In this analysis, it will be assumed that present location theory has at least two distinctive traditions.

The first tradition dates from Alfred Weber[9] himself and perhaps beyond — that of the creation of normative economic models. The approach of such theorists was essentially deductive and based on the operation of a rational economic man. The writers within this tradition were fundamentally interested in creating models with limited variables, but variables with which the hypothetical entrepreneur was completely familiar. Then, in order to test the predictive and explanatory value of the model, the model was compared and contrasted to reality. If the model had a close correspondence with reality, one concluded that in that given instance the entrepreneurs involved were basically motivated by the variables of the model. Alternatively, and in terms of a Darwinian 'survival of the fittest' analogy, the firms which had survived progressive competition were those which were best located as regards the variables of the model. Thus, such writers were mainly interested in the principles which *should* affect the location of industry and how they operated.

The second tradition is of more recent origin and takes a different point of departure — indeed a point of departure that is necessarily critical of the first tradition. The starting point of the behavioural approach is that the focus of location theory should not be on what the entrepreneur should do, but what in fact he does and why he does it. The critical factor behind what the entrepreneur does and why he does it is his access to information. All entrepreneurs acquire information selectively and interpret this information in terms of their own coding mechanism.[10]

These two approaches to the industrial location problem naturally have very different foci. The normative economic tradition has generally analysed the industrial location problem within a framework of perfect competition. The focus has been on the influence of spatially variable costs (transport and production) and economies of scale on location. Meanwhile, the behavioural approach has been principally concerned with the process of decision-making by entrepreneurs or 'technocrats' within corporations on industrial location. In the following attempt at a synthesis of location theory and industrial centralisation in Latin America, these two traditions will be treated separately.

The Normative Economic Tradition

The tradition of normative economic models in location theory stretches back to Weber's classic work and his attempt to explain the location of industry with reference to cost factors. The assumption was that any industrialist would need to take serious note of the spatial variation of cost factors particularly those of transport and labour.

In analysing the utility of Weberian least-transport-cost location theory, Kennelly[11] demonstrated the importance of using real transport networks in the analysis. The most significant generalisation concerning transport networks in less developed countries has been that of Taafe, Morrill and Gould[12] in which they set up an 'ideal-typical' sequence of town and transport growth. The major conclusion concerned the pivotal role of the major port/major city in the evolution of transport networks in less developed countries. Ample evidence is available for modified versions of the model in Latin America — the pivotal role of Buenos Aires in Argentina, Montevideo in Uruguay, São Paulo, Santos and Rio in Brazil, Santiago-Valparaíso in Chile, Lima-Callao in Peru, Guayaquil in Ecuador, and La Guaira-Caracas in Venezuela.

Therefore, the most accessible point in most less developed countries is not the geographic centre but a point on the coastal edge. In this way, when the country began to industrialise, the large coastal city (or city at the centre of the nation's transport system) provided the most advantageous location for the development of three significant industrial types:

1. Consumer-good industries that supply the whole of the national market from one or two plants.
2. Industries in which various raw materials from the hinterland of a country are processed and combined together in one place. This is because the primate city will be the only city

linked to the various sources of the raw material.

3. Industries that have an element of imported raw materials and/or components in their productive structure.

These industrial types have been particularly prominent in the process of import substitution industrialisation (ISI). The original idea of ISI was to substitute imports of a product through the creation of a small number of national plants (Type 1). As the process of ISI continued, plants producing consumer durable products were established. These plants needed to import machinery and components, at least initially, for their manufacturing process (Type 3). Later the manufacture of intermediate products was encouraged, combining either raw materials from the country's hinterland or from abroad (Types 2 and 3).

At the same time, the major metropolitan city provides the optimal location for industries whose major market is that city. In order to identify the importance of this type of industry, it is useful to assess the market size of the principal towns of Latin America. In terms of population alone, the concentration within the primate city can be very high (see Table 6.1). Five countries, (Mexico, Panama, Argentina, Uruguay and Chile) have over 20 per cent of their national population concentrated in the primate city. A further five countries (Costa Rica, Nicaragua, Paraguay, Peru and Venezuela) have between 15 and 20 per cent of their population thus concentrated.

Of course, population concentrations cannot be directly compared with levels of industrial demand, given that the income levels of the populations concerned radically affects that demand. However, it is a well documented fact in Latin America that high-income groups are attracted to the major city in Latin American countries and that as a result the average income per family is much larger in these cities than in smaller cities. Ternent[13] cites data from Mexico where 500,000-plus cities recorded virtually double the monthly family income level of towns with populations between 10,000 and 15,000 inhabitents (see Table 6.2). This is not to deny that income distribution is considerably skewed in the large city. Rather it emphasises that the concentration of demand in the big cities is made up of high average incomes as well as large populations.

It could be argued that although incomes are higher in the big cities, costs are also higher — a factor that would serve to reduce industrial demand. Unfortunately there are very little data on the

Table 6.1: Primacy in Latin America

Country	A Population of primate city (millions)	B Population of country (millions) (1979)	A as % of B	C Population of second city (millions)	A/C
Uruguay	1.17 (1975)	2.9	40.4		
Argentina	9,75 (1978)	26.7	36.5	0.81 (1978)	2.9
Chile	3.69 (1978)	11.0	33.5	0.62 (1978)	6.0
Panama	0.44 (1978)	1.9	23.2	0.15 (1978)	2.9
Mexico	13.99 (1978)	67.7	20.7	2.34 (1978)	6.0
Peru	3.30 (1972)	17.3	19.1	0.30 (1972)	11.0
Venezuela	2.58 (1976)	13.5	19.1	0.79 (1976)	3.3
Paraguay	0.57 (1974)	3.0	19.0		
Costa Rica	0.40 (1977)	2.2	18.2		
Nicaragua	0.40 (1971)	2.5	16.0		
Ecuador	1.02 (1978)	8.0	12.8	0.74 (1978)	1.4
Bolivia	0.66 (1978)	5.2	12.7	0.24 (1976)	2.8
Colombia	2.86 (1973)	26.1	11.0	1.16 (1973)	2.5
Guatemala	0.70 (1973)	6.8	10.3		
Honduras	0.30 (1974)	3.1	9.7	0.15 (1974)	2.0
El Salvador	0.34 (1971)	4.5	7.6		
Brazil	7.20 (1975)	118.7	6.1	4.86 (1975)	1.5

Date of urban census in brackets
Sources: Population Reference Bureau; 1979 World Population Data Sheet; United Nations Demographic Yearbook 1978

Table 6.2: Size of City and Income Levels in Mexico

Size of City (inhabitants)	Monthly Family Income (pesos)
Less than 2,500 (rural)	740
2,501– 10,000	1,000
10,001–150,000	1,450
150,001–500,000	1,900
500,000 plus	2,800

Source: J. A. S. Ternent (1976)

relationship between city size and costs in Latin America. It would appear that housing costs are higher in the bigger cities and, due to greater journey-to-work distances, so are transport costs. However, an interesting retail survey at the end of the 1960s in Chile demonstrated that retail costs were not necessarily higher in the big cities than in cities further down the urban hierarchy. Indeed the

Table 6.3: The Relationship Between City-Size and Retail Prices, Chile, 1970

City	City-size	Indices of retail sectors[2] 1	2	3	4	5	Index total	Totals indexed[3]	Average index for each city-size range[4]
Santiago[1]	2,860,000	138	100	111	102	136	587	110	110
Concepción[1]	540,000	132	132	123	111	136	634	119	118
Valparaíso[1]	530,000	157	129	119	108	122	625	117	
Antofagasta	125,000	144	157	143	105	124	673	126	
La Serena/ Coquimbo	116,000	143	110	131	103	138	625	117	116
Temuco	113,000	134	118	114	106	128	600	112	
Rancagua/ Machalí	109,000	137	111	108	102	121	579	108	
Talca	94,000	134	102	121	107	151	615	115	
Chillán	88,000	127	117	115	101	118	578	108	
Valdivia	83,000	126	132	119	107	129	613	115	113
Iquique	65,000	128	185	136	107	105	661	124	
Puerto Montt	64,000	100	124	100	102	113	539	101	
Los Angeles	49,000	118	103	111	100	102	534	100	
Curicó	42,000	134	102	118	103	131	588	110	
Linares	38,000	134	132	136	112	118	632	118	109
San Fernando	30,000	132	110	119	111	100	572	107	
Angol	27,000	117	128	107	102	123	577	108	
Cauquenes	21,000	125	121	133	104	108	591	111	

1. The metropolitan zone
2. Price totals for each retail sector were derived by adding the average retail prices (1970) of related goods as follows:
   1. **Meat:** Summation of average prices for seven common cuts of meat (1 kg of each)
   2. **Salad vegetables and fruit:** Summation of average prices for: 1 kg of onions, potatoes, apples, oranges; a bunch of carrots; a cabbage and a lettuce
   3. **Dairy products:** Summation of average prices for 10 pt. of milk, 1 kg of butter, 1 kg of cheese, one tin of condensed milk and 1 tin of powdered milk
   4. **Basic foodstuffs:** Summation of average prices for: 2 kg of rice, flour and sugar; 1 kg of beans, spaghetti, jam and salt; 0.5 kg of tea; 170 g of coffee; 250 g of cornflour; 1 litre of cooking oil; 10 eggs
   5. **Textiles and clothing:** Summation of average prices for 1 blanket, 2 sheets, 2 towels, 2 men's shirts, 2 vests, pair of shoes (men's), 10 m of cotton fabric

   The resultant price totals were then indexed with the base (100) being the lowest price total of the particular retail sector
3. Base = 100 = lowest index total = Los Angeles = 534
4. The city-size range is as follows:
   Primate city (2,860,000); Major metropolitan zones (500,000–600,000); Major regional centres (100,000–130,000); Major provincial centres (60,000–100,000); Provincial centres (20,000–60,000)

Source: Instituto Nacional de Estadisticas (1971) *Comercio Interior y Servicios, Año 1970* (Santiago)

relationship between city size and the retail price index could be described as an inverted U-shaped curve, with the lowest average price indices occurring at both ends of the city-size range[14] (see Table 6.3). In terms of the primate city, this was probably due to the establishment of new concepts in retailing (such as the hypermarket and supermarket) and the consequent increases in productivity and lower per unit margins that resulted. Furthermore, marketing institutions in the primate city often had the benefit of cheap transport rates for products coming from the agricultural provinces. Cities further down the hierarchy were not benefiting from the cost benefits of such institutions. Although such evidence is only partial to the overall cost structure of large cities, it does imply that the higher relative incomes enjoyed in the larger cities are not necessarily reduced in comparative terms by higher costs.

According to Weber, cheaper labour costs at other locations can attract entrepreneurs away from the point of minimum transport costs. It has been presumed that labour costs in the primate city are higher than the national average and considerably in excess of those prevalent in the peripheries of Latin American countries. However, a comparison of the Chilean 1967 Industrial Census (all firms over 5 employees) and the 1978 Industrial Survey (all firms over 50 employees) reveals a different spatial distribution and structure of labour costs (see Table 6.4 and Figures 6.1, 6.2, 6.3 and 6.4). In terms of both white- and blue-collar workers, the Chilean primate city of Santiago had both average salaries and wage rates below or near the national average on both occasions. It should be borne in mind that between 1967 and 1978, Chilean industry underwent major structural changes, first undergoing Marxist policies of state control[15] and then experiencing monetarist policies stressing the comparative advantages of the resources that Chile possessed. Despite such major shifts in industrial policy and direction during the period, the pattern of provincial ranking of industrial salaries and wages remained remarkably stable.

Those provinces dominated by export industries tended to have salary and wage rates higher than that of the primate city. The average ranking of the copper-exporting provinces of O'Higgins, Atacama, and Antofagasta in terms of both white- and blue-collar workers was high in both years, while the timber- and cellulose-exporting provinces of Bío-Bío, Arauco and Ñuble improved their combined average ranking substantially during the period (from 13 to 6 in terms of white-collar workers). As export industries are

Table 6.4: The Changing Rank of Chilean Provinces Between 1967 and 1978 in Terms of Salaries and Wages

Province	Ranking in terms of provincial average for white-collar salaries in manufacturing		Ranking in terms of provincial average for blue collar wages in manufacturing	
	1967	1978	1967	1978
Tarapaća	9	4	6	3
Antofagasta	6	5	2	6
Atacama	1	12	1	2
Coquimbo	19	16	9	18
Aconcagua	17	15	14	13
Valparaíso	5	6	4	5
Santiago	7	7	10	8
O'Higgins	2	1	7	1
Colchagua	10	8	11	7
Curicó	22	14	21	20
Talca	18	10	13	9
Linares	8	21	15	12
Ñuble	15	13	19	10
Concepción	4	9	3	4
Arauco	21	2	24	14
Bío-Bío	3	3	5	15
Malleco	13	17	20	21
Cautín	20	22	18	22
Valdivia	16	18	17	16
Osorno	11	11	12	11
Llanquihue	14	19	16	17
Chiloé	23	24	22	24
Aysén	24	23	23	23
Magallanes	12	20	8	19

Sources: Instituto Nacional de Estadisticas (1971) — IV Censo Nacional de Manufacturas, 1967; Instituto Nacional de Estadisticas (1981) — Industrias Manufactureras: Año 1978

linked to world rather than national markets, there is greater elasticity in wage and salary rates in regions dominated by such industries. However, more significant for the location of consumer good industries is the fact that Chile's second and third conurbations of Valparaíso and Concepción generally had higher average salaries and wages in manufacturing than Santiago on both occasions (the exception was Concepción's average salaries in 1978). The major decentralisation schemes of Chile, located in Arica and Iquique in the province of Tarapacá, also had higher salaries and wages than the primate city. Thus, in Chile, the most

Figure 6.1: Provincial Variation in Annual Average Salaries for White Collar Workers in Chile 1967

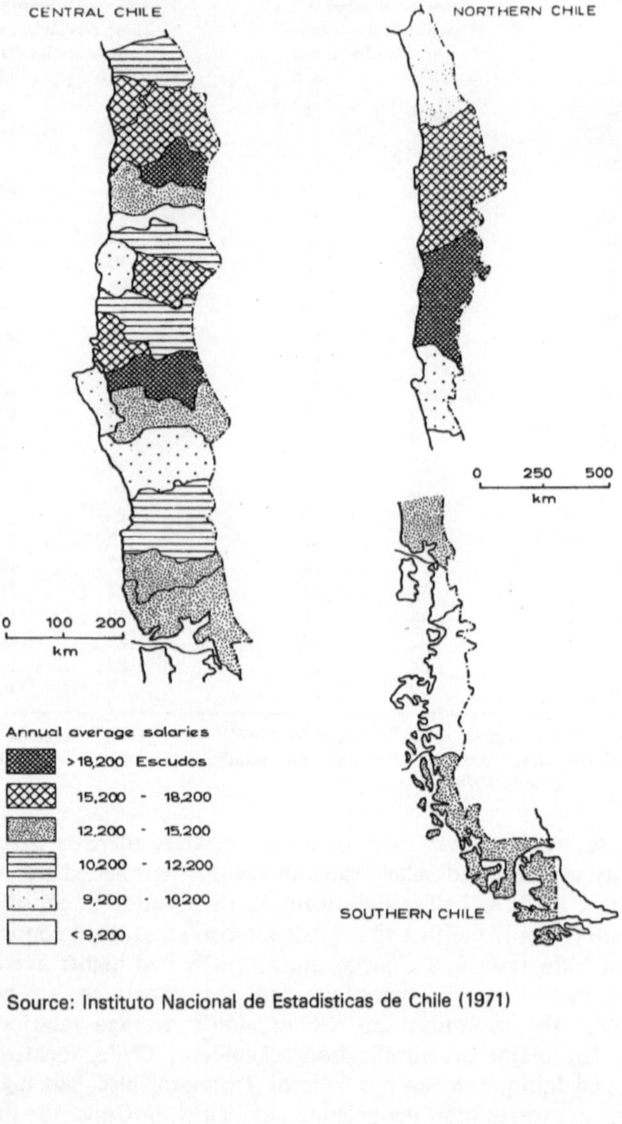

Source: Instituto Nacional de Estadisticas de Chile (1971)

Figure 6.2: Provincial Variation in Annual Average Wage Rates for Blue Collar Workers in Chile 1967

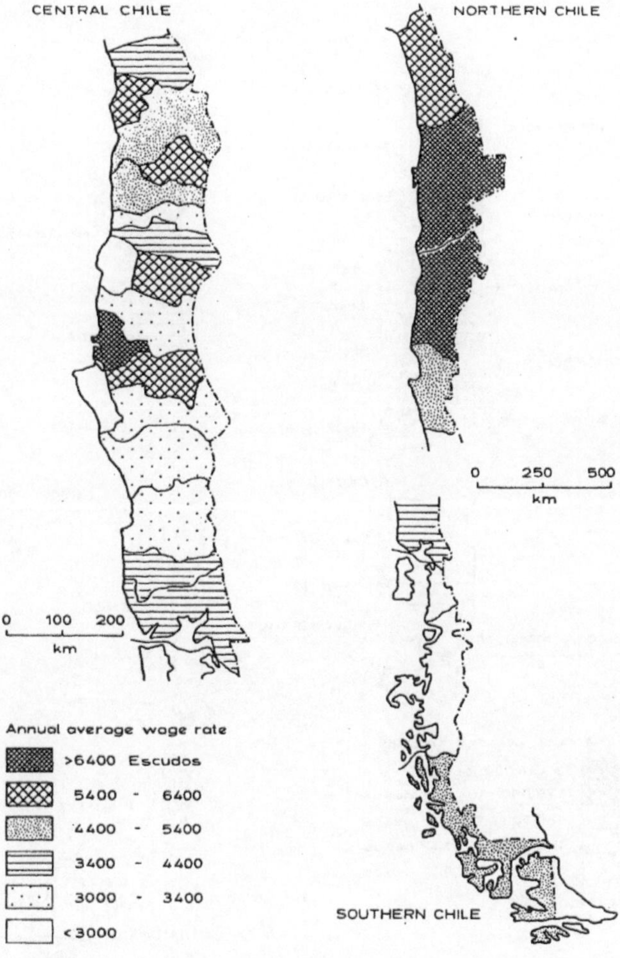

Source: Instituto Nacional de Estadisticas de Chile (1971)

Figure 6.3: Provincial Variation in Annual Average Salaries for White Collar Workers in Chile 1978

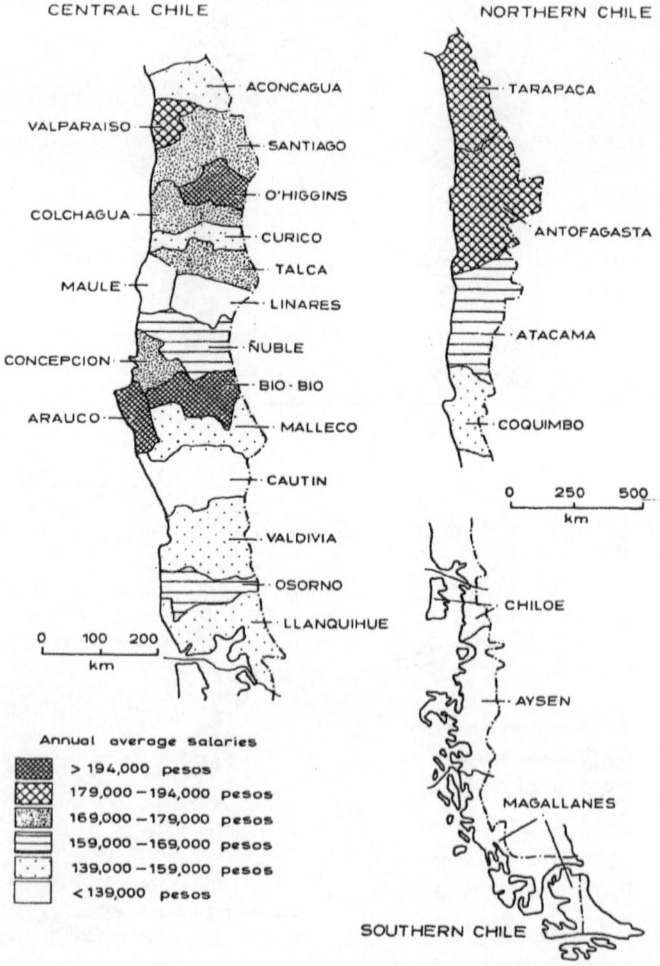

Source: Instituto Nacional de Estadisticas (1981) Industrias Manufactureras: Año: 1978

Figure 6.4: Provincial Variation in Annual Average Wage Rates for Blue Collar Workers in Chile 1978

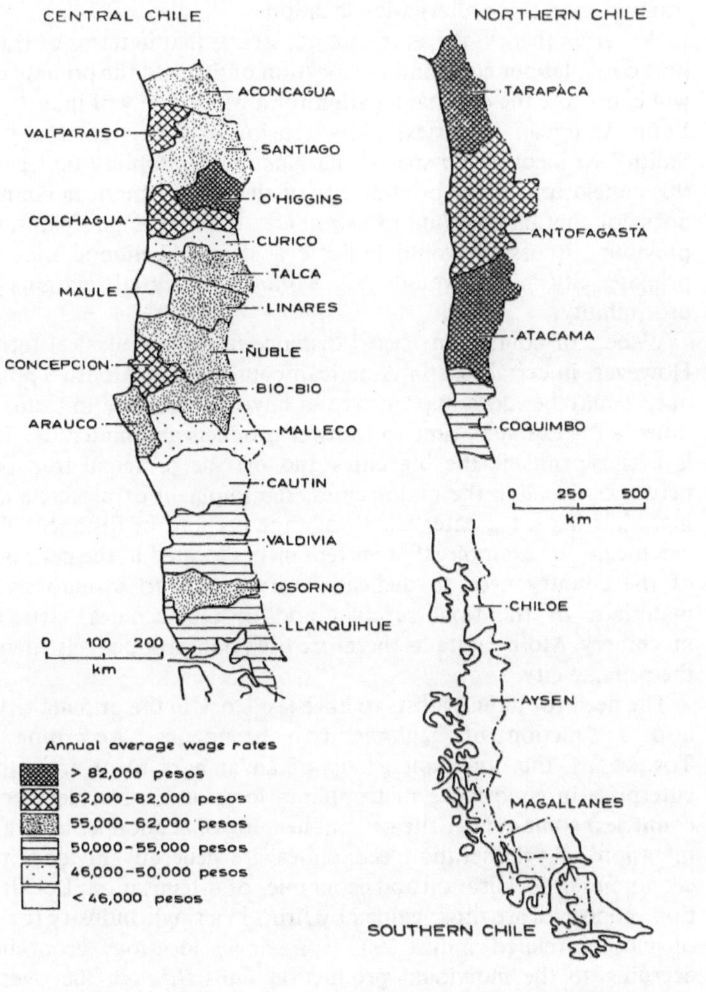

Source: Instituto Nacional de Estadisticas (1981) Industrias Manufactureras: Año 1978

attractive alternative locations for industrial entrepreneurs had higher average labour costs than those prevalent in Santiago. Meanwhile, Santiago could offer a wider range of labour skills and training than these alternative locations.

Weberian theory, therefore, demonstrates that in terms of transport costs, labour costs and the location of demand the primate city will constitute the optimal location for a wide variety of industry in Latin American countries, The extension of Weberian theory by Smith[16] to incorporate spatial margins of profitability underlines this conclusion. If the primate city of the Latin American country does not contain the point of lowest cost for a particular plant, the previous discussion would indicate a strong likelihood that the primate city plant would fall *within* the spatial margins of profitability.

Weber, of course, envisaged distance in simple physical terms. However, in certain Latin American countries, as Alonso[17] points out, it may be more appropriate to envisage distance in terms of time — for example, time to transfer products and materials. This is because outside the big cities and off the principal transport networks that link the major cities, the shipment of products and materials takes a greater length of time per unit of distance. This can mean, for example, that entrepreneurs located in the periphery of the country need to dedicate more capital to inventories of materials, of the final product and of replacement parts for machinery. More capital is therefore used non-productively than in the primate city.

The need for industrialists to have less stock in the primate city is also a function of agglomeration economies. According to Townroe,[18] 'the dominant group of advantages to an industrial enterprise in choosing a metropolitan location (in less developed countries) comes under the general heading of economies of agglomeration'. Agglomeration economies are generally divided into economies of localisation and economies of urbanisation. Localisation economies are those gained by firms in a single industry (or set of closely related industries) at a single location, economies accruing to the individual production units through the overall enlarged output of the industry as a whole at that location. Urbanisation economies apply to all firms in all industries at a single location and represent those external economies passed on to enterprises as a result of savings from the large-scale operation of the agglomeration as a whole.

Localisation economies may at first glance appear to have limited applicability in large Latin American cities. However, there is growing evidence of the spatial concentration of certain industries in certain cities — such as the Brazilian motor vehicle industry in the Greater São Paulo conurbation. Furthermore there is growing evidence of sub-contracting[19] linkages within industries in large cities. Roberts[20] refers to complex networks of production in which a modern factory may commission a small, family enterprise to undertake part of its production and Scott[21] provides examples of these networks in such manufacturing sectors as the assembly of refrigerators, the upholstery of microbuses and dressmaking in Lima. Even when the modern factory is located in a peripheral region, such sub-contracting linkages will be concentrated in the primate city — as with the early development of the Chilean motor vehicle industry.[22] In 1968, when the majority of Chilean motor vehicles were assembled in the northern port of Arica and when over 50 per cent of parts by value had to be Chilean made, 89 per cent of the 300 parts-producing plants that had been created in Chile were nevertheless located in Santiago; only 8 per cent of the plants had decided to locate near the assembly plants in Arica. The metal-working and other small firms that ventured into components production in Santiago were generally not new firms but existing companies that were intent on adding a further line of production to their organisation. The concentration of this small-scale manufacturing economy in the primate city is one of the principal elements of localisation economies in Latin America and can effectively reduce costs for the large-scale firm located there.

Urbanisation economies can also have powerful attractions for industrial decision-makers. The effect of economies of massed reserves and large-scale purchasing is to reduce the unit cost of inputs and reduce the amounts of non-productive capital in the production process in comparison with similar cities.[23] Furthermore, the nature of urban systems in Latin America provides the primate city with a greater variety and quality of services. An example of this is banking. In Chile, all major financial decisions made by the provincial branches have to be referred to the bank's head office in Santiago for ratification, entrepreneurs in the capital will therefore have a much more rapid and often more favourable access to capital than those in the provinces. Similar concentration of banking functions is found in Mexico where the Federal District accounts for 68 per cent of the total capital stock and reserves in the national banking system and 93 per cent of the long-term deposits. Furthermore, the banks of the Federal District grant 76 per cent of the national total of mortgage loans and handle 68 per cent of the country's investments in stocks and bonds.[24]

Finally the infrastructure of a primate city, with an international airport, central railway system and spare power capacity has many advantages over the comparable infrastructure of smaller cities. In many Latin American countries, there have been cases of small towns lacking the necessary power capacity to cope with the operation of new high-energy industries at peak times; the consequent stoppages can have severe cost penalties for industries using continuous casting or other such processes. An example from Chile is interesting in this respect. In 1969, the Chilean company, Manufacturas de Cobre (MADECO), decided to locate a plant making copper telephone cables in the northern port of Antofagasta. Despite the fact that Antofagasta was the fourth city of Chile and despite its proximity to the Chuquicamata copper mine, an enormous user of electricity, the MADECO plant suffered for seven years from cuts in electricity supply at peak periods of demand. As MADECO had smelters which used a continuous casting process, where temperature is kept at a constant to produce the required type of refined copper, a cut in electricity meant the total loss of the copper being refined and a period of no production as the smelters cooled, were cleaned out and prepared again for production.

However, it has frequently been pointed out that the considerable infrastructure that the capital city possesses has normally been heavily subsidised by national funds[25] and that the entrepreneur in the primate city does not pay the real costs for such services. For example, in Mexico, the price of electricity is the same in Mexico City as in all other parts of the country despite the fact that the bulk of electricity consumed by the city is generated at the hydro-electric complexes of Malpase and Infiernillo, 1000 and 600 kilometres away respectively.[26] The considerable cost of transmission is not directly passed to the entrepreneurial consumer.

The normative economic tradition in location theory has strongly emphasised that entrepreneurs should be influenced in their locational decisions by the spatial incidence of cost factors — notably transport costs, costs of serving the main markets, labour costs and the cost benefits of agglomeration economies. When such a framework is applied to the spatial pattern of industrial development in

Latin American countries, it achieves considerable power in explaining the centralisation of industry in those countries and the strong forces that are still promoting such centralisation.

The Behavioural Problem

A problem of the normative economic tradition was the assumption within the locational models of the operation of rational economic man. The behavioural tradition in locational analysis starts with the very different assumption that when an entrepreneur makes a decision with spatial implications (plant expansion, plant closure, plant reduction, movement of plant, creation of a branch plant) the decision is made *not* according to economic models but rather in terms of information that enters into his consciousness and in function of his attitude towards that information — information that is both incomplete and biased. The behavioural tradition has therefore fostered numerous empirical analyses of decision-making but, unlike the normative tradition, has been unable to develop many generalisations of locational behaviour.

One generalisation that has been developed with reference to less developed countries[27] is that the entrepreneur prefers to make his initial decision such as the establishment of a plant, in an environment in which he feels secure. This could indicate a great stimulus to the decentralisation of industry if the entrepreneur established his initial undertaking in his home town where he could readily make contacts on both the supply and demand sides of his operation. However, according to the majority of migration studies in Latin America,[28] the more enterprising entrepreneurial members of provincial populations migrate to the major cities of the country at an early age. If and when the entrepreneurial migrant adapts to city life, it would be in a major city that he would feel most secure to set up a business. As a consequence, most censuses of big Latin American cities reflect steady growth in the 'self-employed' category.[29]

For the successful entrepreneur in the large-scale sector of the national economy, the primate city presents an attraction for head office location, if not for plant location, due to what Tornquist[30] has termed the entrepreneur's contact patterns. According to Tornquist, face-to-face contacts are becoming increasingly important to upper level business and government administrators. This is particularly appropriate in Latin America where business 'procedures follow less standardised forms' and where 'business transactions tend to be more of a social phenomenon' than in the USA.[31] Such personal contacts are best facilitated within the framework of large urban centres. In Mexico, for example, 29,185 (or 56 per cent) of the country's 51,619 members of industrialists' organisations lived in the Federal District and state of Mexico in 1975.[32]

Increasingly manufacturing activity in Latin America is being organised through large corporations, whether these be of state, private domestic or international origin. The largest and most powerful corporations are normally those linked to the state, such as PEMEX, PETROVEN, SIDERBRAS or CODELCO. Their head offices are firmly located in the capital city, which with the notable exception of Brazil, normally duplicates as the primate city. More importantly the investment of state firms can be highly centralised. For example, Brazilian state investment in infrastructure and heavy industry has been largely organised through the National Development Bank (BNDE). However the regional distribution of the Bank's investment has shown a marked spatial concentration in the Centre-South region.[33] Between 1952 and 1965 the region received 96.6 per cent of all BNDE investment in manufacturing. Although this exaggerated level of spatial concentration in state investment has declined subsequently, the Centre-South still accounted for over 60 per cent of BNDE manufacturing investment in 1974. Meanwhile, the Industrial Development Commission, an agency set up by the government to encourage private sector investment in priority industries through tax and depreciation allowances, had an even more spatially concentrated distribution of investment. In 1975, the Centre-South region of Brazil, including São Paulo, Rio and Belo Horizonte, accounted for nearly 80 per cent of all industrial investment approved by the CDI.[34]

In the two recent decades, there has also been a well-defined process of organisational concentration among the leading national private firms of Latin American countries — such firms as the Mendoza group and SIVENSA in Venezuela. In Chile, the process has been particularly marked due to the radical restructuring of that country's industrial economy between 1974 and 1978; by the end of 1978, six Chilean firms controlled two thirds of total assets in Chile (see Chapter Four). The head offices of these six large conglomerates were in downtown Santiago despite the fact that one of the conglomerates (the Angelini) has the majority of its productive plant linked to the export economy of peripheral regions (see Table 4.1).

Multinational corporations are the third element of the increasing corporate nature of manufacturing activity in Latin America. In certain industries, there is a strong emphasis on linking plants between countries. In terms of the motor vehicle industry, Volkswagen is trying to integrate its Brazilian and newly-acquired Argentine operations in order to maximise economies of scale. In these cases, the primate city, with its privileged international transport links, becomes the focus for most industrial expansion. Furthermore, behavioural analyses of locational decision-making abroad have found that executives of multi-national corporations favour locations in cosmopolitan cities over cities further down the urban hierarchy. These preferences may be formalised as in the case of IBM's decision-making framework[36] where 90 per cent of the points allocated tend to favour the primate city in Latin American countries (living conditions, accessibility, level of industrialisation, labour availability, prestige effect of the location, community attitude).

Multinational corporations tend to operate at two different levels in many Latin American countries.[37] On one level, there is the productive plant which will be located according to the pull of such factors as manpower, raw materials and markets. If the raw material pull is important or if that of labour has some significance, as in the case of a primate city with labour shortages or higher labour costs, a decentralised location may result. But the other level at which multinational corporations operate in Latin American countries, that of the regional office, organisationally linking the productive activities in the country with head office and specialising in sales, administration, public relations, accountancy and legal affairs, is invariably located in the primate city. Such a location has the major advantages of accessibility and the easy transfer of executives from head office to regional office. An appropriate example of this is provided by the Venezuelan motor vehicle industry, dominated by international companies. With the exception of General Motors, these companies have all located their productive plants outside Caracas, mainly in the axial belt that runs from Guatire in the east to Morón in the west (see Figure 7.3). However, without exception, all regional offices are located in the Caracas basin. The need to be near the appropriate government body, the Transport Equipment Division of the Development Ministry, and the major international airport, Maiquetia, were the two principal reasons given in a survey of the industry in 1979.[38] As in most examples of government-controlled import substitution, the Venezuelan Transport Equipment Division tightly controls the development of both the assembly and the components industry through establishing the parameters within which they must grow. The decentralised location of the Venezuelan motor vehicle industry demonstrates that even if the production plant has been established outside of the capital city, it is still likely that the regional office of the international company will be located in the capital. With improvements in international telecommunications, office technology and computer services being spatially concentrated in the major city, such a pattern can only become more exaggerated in the future.

The behavioural approach, by focusing attention on the factors in the process of locational decision-making, demonstrates that the primate city is a powerful attraction for manufacturing entrepreneurs of all kinds — both in the small-scale and large-scale sectors, in state and private sectors and in national and multinational sectors. The need for interaction with other entrepreneurs from various sectors of the urban industrial economy and the increasing need for face-to-face contacts at the upper levels of management and in relation to government have produced coding mechanisms in which a great majority of entrepreneurs seek information about and decide to locate in the primate city.

Location theory can thus contribute to the explanation of patterns of industrial centralisation in Latin American countries and the processes behind these patterns. The normative economic tradition was used to explore the primate city as a low-cost location for a wide variety of industrial sectors. The behavioural approach was able to demonstrate the attraction of the big city for decision-makers in the ever-increasing corporate sector of Latin American economies.

However, an important reason for applying location theory to a generalised Latin American spatial framework has been to identify and dissect some of the complex processes behind the centralisation of industry. In the aggregate, the operation of complex and wide-ranging processes leads one to conclude that the strength of centralising forces is substantial and to predict that such spatial centralisation can only become more accentuated in the future. The analysis showed that forces promoting the decentralisation of industry are relatively weak. Therefore, unless the spatial centralisation of industry and the concomitant concentration of employment and population is welcomed, some intervention or initiative of government may be necessary. In this context, it is instructive to analyse the theory behind and the empirical record of recent attempts at industrial decentralisation in Latin America.

References

1. J. Dickenson, 1978, *Brazil*. (Dawson, Folkestone)
2. S. Levi de Lopez, 1980, Industrial Development in Mexico, *Bulletin of the Society for Latin American Studies*, 32, p. 14
3. A. G. Gilbert, 1974, *Latin American development: a geographical perspective*. (Penguin, Harmondsworth)
4. Dirección General de Estadistica y Censos Nacionales, *V Encuesta Industrial 1975: Resultados Regionales*. (Ministerio de Fomento, Caracas, 1977)
5. A. G. Gilbert, *Latin American development*
6. R. N. Gwynne, 1978, Industrial Decentralisation in Chile. (Unpublished doctoral dissertation, University of Liverpool)
7. A. Weber, 1929, *Theory of the Location of Industries*. (University of Chicago Press, Chicago)
8. P. E. Lloyd and P. Dicken, 1977, *Location in Space: A Theoretical Approach to Economic Geography*. (Harper and Row, London)
9. A. Weber, *Theory of the Location of Industries*
10. P. Dicken, 1971, Some Aspects of the Decision-Making Behaviour of Business Organisation, *Economic Geography*, 47, pp. 426–37
11. R. A. Kennelly, 1968, The Location of the Mexican Steel Industry, in R. H. T. Smith, E. J. Taafe and L. J. King (eds.), *Readings in Economic Geography*. (Rand McNally, Skokie, Ill.), pp. 126–57
12. E. J. Taafe, R. L. Morrill and P. R. Gould, 1963, Transport Expansion in Underdeveloped Countries: A Comparative Analysis, *Geographical Review*, 53, pp. 503–29
13. J. A. S. Ternent, 1976, Urban Concentration and Dispersal: Urban Policies in Latin America, in A. Gilbert (ed.), *Development Planning and Spatial Structure*. (John Wiley, London), pp. 169–85
14. R. N. Gwynne, 1978, City Size and Retail Prices in Less Developed Countries: an insight into primacy, *Area*, 10, pp. 136–41
15. R. N. Gwynne, 1976, *Economic Development and Structural Change: The Chilean Case, 1970–1973*. (Department of Geography, University of Birmingham, Occasional Publication No. 2)
16. D. M. Smith, *Industrial Location*. (John Wiley, New York, 1971)
17. W. Alonso, 1968, *Industrial Location and Regional Policy in Economic Development*, Center for Planning and Development Research, University of California, Berkeley, Working Paper No. 74
18. P. M. Townroe, 1979, Employment Decentralization: Policy Instruments for Large Cities in Less Developed Countries, *Progress in Planning*, 10, pp. 85–154
19. J. Dickenson, *Brazil*
20. B. Roberts, 1978, *Cities of Peasants*. (Edward Arnold, London)
21. A. M. Scott, 1976, Who are the Self-Employed? in C. Gerry and R. Bromley (eds.), *The Casual Poor in Third World Cities*. (John Wiley, London)
22. R. N. Gwynne, 1978, Industrial Development in the Periphery: The Motor Vehicle Industry in Chile, *Bulletin of the Society for Latin American Studies*, 29, p. 47

23. W. Alonso, *Industrial Location and Regional Policy in Economic Development*

24. G. Garza and M. Schteingart, 1978, Mexico City: The Emerging Metropolis, *Latin American Urban Research*, 6, pp. 51–86

25. A. G. Gilbert, 1976, The Arguments for Very Large Cities Reconsidered, *Urban Studies*, 13, pp. 27–34

26. G. Garza and M. Schteingart, Mexico City

27. J. O. C. Onyemelukwe, 1974, Industrial Location in Nigeria, in F. E. I. Hamilton (ed.), *Spatial Perspectives on Industrial Organisation and Decision-making*. (John Wiley, London), pp. 461–84

28. B. Roberts, *Cities of Peasants*

29. Ibid.

30. G. Tornquist, 1970, Contact Systems and Regional Development, *Lund Studies in Geography*, Series B35

31. W. Alonso, *Industrial Location and Regional Policy in Economic Development*

32. G. Garza and M. Schteingart, Mexico City

33. J. Dickenson, *Brazil*

34. Ibid.

35. F. Dahse, 1979, *El mapa de la extrema riqueza* (Editorial Aconceagua, Santiago)

36. A. Blackbourn, 1974, The Spatial Behaviour of American Firms in Western Europe, in F. E. I. Hamilton (ed.), *Spatial Perspectives on Industrial Organisation and Decision-making*. (John Wiley, London), pp. 254–64

37. See S. Hymer, 1975, The Multinational Corporation and the Law of Uneven Development, in H. Radice (ed.), *International Firms and Modern Imperialism*. (Penguin, Harmondsworth), pp. 37–62

38. R. N. Gwynne, 1979, The Venezuelan automobile industry, *Business Venezuela*, 64, pp. 24–38

7 PROCESSES OF INDUSTRIAL DECENTRALISATION

Industrial centralisation can be seen as both a pattern and process of the industrial geography of Latin America. With reference to total industry and individual industrial sectors, centralisation is a pattern — the spatial pattern of industries in aggregate concentrating in the primate city. However, with reference to the decisions taken by the entrepreneurs and executives of individual firms, centralisation can also be seen as a process. This reflects the emphasis of locational decisions taken in state, multinational and private domestic firms to create and expand plant in the primate city and to ignore or reduce plant in the periphery of Latin American countries.

The pattern of industrial centralisation is becoming more and more pronounced. Two factors help to explain this. First, the technologically advanced and rapidly growing new industries are consistently drawn to the primate city. For example, four sectors of Chilean industry in 1967, pharmaceuticals, electrical durable goods, plastics and professional and scientific equipment had over 90 per cent of their employment and value-added generated in Santiago.[1] Secondly, very large concentrations of population are being created in the metropolitan areas and outer agglomerations of the primate cities. For example, in 1984, the agglomeration of Mexico City had an estimated population of sixteen million and São Paulo of twelve million. Both are increasing their populations by an estimated 600,000 people per annum. With such massive, rapidly growing populations, market-oriented industries of all types are strongly attracted to the primate city location.

The merits and demerits of industrial centralisation are closely linked to the wider debate about the value of large agglomerations for national and regional development in Latin America.[2] More specifically, it can be demonstrated that the large city offers the locational attraction of a wide range of industrial types within a relatively small area. This gives the entrepreneur the advantage of being able to build up in a geographically concentrated area a dense network of material linkages (with suppliers and stockists) and market linkages (with consumers, whether companies or individual purchasers). The entrepreneur benefits from a wide range of labour skills and a relatively high standard of infrastructure — particularly in terms of communications and power. Furthermore, there is no need to invest in social capital — housing, health and education. While the advantages for the individual entrepreneur or executive are significant, the wider issue of national and regional development introduces a new perspective. For the corollary of such industrial centralisation is that industrial growth, technological change and associated impulses for innovation tend to bypass or have little impact on the economic growth of those regions and towns away from the primate city. In so doing, individual Latin American countries become characterised by the contrast between a dynamic and innovative centre and a stagnant and weak periphery. Such can be the perception of businessmen and entrepreneurs of the phenomenon that comments such as 'Santiago is Chile' sum up this contrast in a more exaggerated way than reality would suggest.

Owing to the wider issues raised by industrial concentration for national and regional development, industrial decentralisation has become a topical issue and policy since the 1950s. As a policy it does require conscious government effort and intervention and a capacity for national and regional planning. This chapter aims to analyse the record of industrial decentralisation policies in Latin America over the last thirty years. It will begin by looking at some of the theories used to promote industrial decentralisation before moving on to examine some general trends and processes that have become evident during the period.

Growth Pole Theory

The theory most prevalent behind schemes of industrial decentralisation in Latin America in the late 1950s, 1960s and early 1970s was that concerning growth poles. The basic growth pole theory was largely developed in France, a country that shares the spatial phenomenon of an accentuated centralisation of industry. Perhaps this was one of the reasons for its wide diffusion and easy acceptance by regional planners in most Latin American countries in the 1960s.

Growth pole theory traces its origins to 1950 with a paper by a noted French regional economist, Perroux.[3] Perroux's research had convinced him that economic theory and research in macroeconomics was too concerned with the problems of individual countries. It was as if the discipline of economics saw each country as a container and tried to analyse economic issues within that container. This led economics to have little interest in the dynamics of international economic growth and to make analysis of the international nature of firms and their growth patterns difficult. Perroux preferred to conceive of the firm as operating within what he termed 'abstract' space. In particular, Perroux defined two forms of abstract economic space for a firm:

1. *Plan space* — the space which defines the relations between the firm on the one hand and its suppliers and buyers on the other. For example, in terms of the motor vehicle firm, the 'plan space' would, in a non-spatial form, refer to all the suppliers to the firm and all the distributors which sold the final product. It was Perroux's contention that this would prove a better way of analysing industrial growth. By identifying a growth firm or a firm with an innovation, the implications for related firms could be better projected.
2. *Space defined as a field of forces*. Forces could either be centripetal (attracting men and materials to the firm) or centrifugal (repelling other activities or competitors).

In the early 1950s, Perroux was involved in research on the Ruhr industrial zone, and perhaps as a consequence, shifted the analysis in a 1955 paper to look at industrial growth in terms of real geographic space.[4] He noted that:

1. Geographic space could be defined as a field of forces as growth does not appear everywhere at the same time but rather at certain points or 'poles'.
2. A growth firm or a firm with an innovation acts as a focus to new industrial development in geographic as well as abstract space. These 'propulsive' industries, in expanding their output, increase the sales and purchases of several other industries which are linked to them as buyer or seller. A propulsive industry will then cause an increase in the sales of a whole group of industries which is very much larger than the increase in its own sales.

These two aspects of Perroux's analysis, originally envisaged in abstract space, became in his later reformulation intertwined with geographic space, and from this has arisen much confusion about the precise nature of growth poles. By the time Boudeville wrote in 1966, the two concepts had become firmly intertwined.[5] His definition was that 'a regional growth pole is a set of expanding industries located in an urban area and inducing further development of economic activity throughout its zone of influence'. Whether it is valid to bring together Perroux's two ideas and merge them into one is debateable. What is true is that many regional economists and planners in Latin America and elsewhere *did* merge the two concepts.

However, the merging of abstract and geographic space in the growth pole concept does have some serious theoretical flaws. The basis of the growth pole concept depends upon the idea of linkages, both vertical and horizontal, between industrial plants. Hirschman pointed to some of the possible problems that can affect linkages between the propulsive and dependent firms when the ideas of Perroux are taken from 'abstract' and placed into 'geographic' space.[6] He recognised that linkages could be considered both in terms of their strength and of their importance. The 'strength' of the linkage refers to the probability that other industries will be set up as a result of the impulse provided by the establishment of the propulsive firm or firms. The 'importance' of the linkage refers to the growth in employment or production that might be stimulated by the establishment of the dynamic sector. The actual multiplier impact, therefore, depends both upon the 'strength' (probability) and the 'importance' (potential employment or production generation). But, as Hirschman indicated, the two are inversely related. Thus, if the linkage is 'strong' it is likely that it will also be small in importance. He illustrated the point by reference to the linkage between a cement producer and a multi-walled paper bag producer. The establishment of the cement producer with requirements for a large volume of packaging material may stimulate the setting up of the bag producer which will sell all of its output to the cement company. The probability of this linkage stimulating the establishment of the second company is high although the 'importance' is limited — in terms of the employment that will be generated. On the other hand, the multi-walled bag producer would constitute a market for a large scale paper manufacturer but not, of course, a market of sufficient size to cause the company to be established. In this case the 'importance' of the linkage is great but its 'strength' limited.

When the effect in geographical space is considered, the strength of any such linkage is likely to be weakened still further because there may be no incentive on the part of the dependent plant or firm to locate in the proximity of the propulsive industry. Continuing Hirschman's example, for instance, there is no inevitability about the multi-walled paper bag producer locating in the same city as the cement plant. The optimum location for such a plant might be at the minimum transport cost point between several cement factories, and so the establishment of a cement plant as a potential growth centre would merely stimulate the output of multi-walled bags already being produced in a large industrial city.

Industrial Decentralisation and the Motor Vehicle Industry

It could be argued that the problems of strength and importance would be lessened if the propulsive industry were carefully selected. Boudeville, for example, stated that the successful propulsive firm should be characterised by three features: a high degree of interaction with other firms, a dominance over other firms' production either through supplying inputs or purchasing outputs and large firm size. Such characteristics should provide for the generation of a considerable number of local multipliers. In the terms of Hirschman, such a firm would provide a large number of highly probable linkages of low importance; however, the net effect should theoretically be substantial.

One type of firm that can lay claim to possessing these three attributes is the motor vehicle firm; in a regional context, it would apply to the vehicle assembly plant. A motor vehicle firm has large size, a certain dominance over the production of related firms mainly through supply and a high degree of interaction with other firms; vehicles are assembled from between 2,000 and 20,000 parts, most of which come from different manufacturing plants. It is for these reasons that the motor vehicle plant has become a popular candidate for regional growth poles — in Europe and the USA as well as Latin America. In Latin America, motor vehicle plants have been encouraged to locate in such decentralised locations as Trujillo in Peru, Barcelona in Venezuela, Cordoba in Argentina, Belo Horizonte in Brazil and Arica in Chile. It is instructive to analyse the latter case as many of the tendencies found here are present in the other examples of decentralisation of motor vehicle plants.

In 1962, the Chilean government decided to concentrate its

Figure 7.1: The Location of the Motor Vehicle Industry in Chile

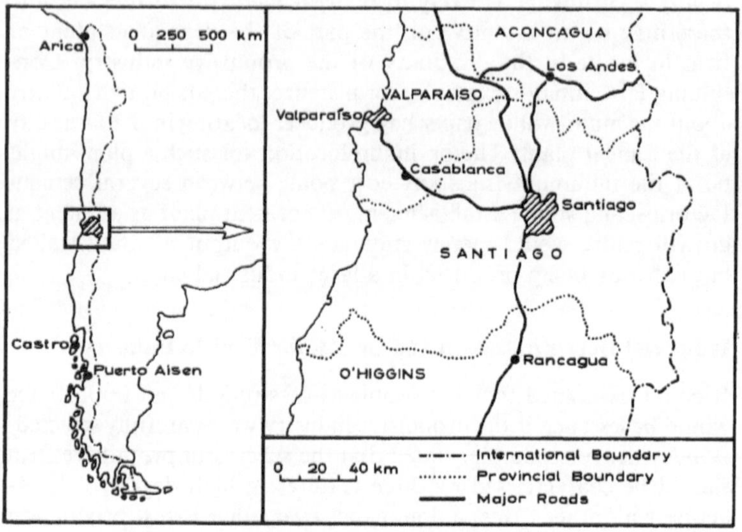

motor vehicle assembly industry in the northern industrial growth pole of Arica, 2,000 kilometres from the capital, Santiago (see Figure 7.1). It hoped along the lines of growth pole theory that a large number of local multipliers from an expanding industry would radically change the urban economy of Arica. A special industrial estate was constructed for this purpose.

However, local multipliers were to be very restricted in number. The basic problem was that as motor vehicle production rose from negligible amounts in 1960 to over 26,000 in 1972 and as local content increased from nothing to 58 per cent at the same time,[7] the multipliers from the expanding industry were mainly felt outside Arica and most notably in the capital, Santiago. On the supply side, one can see this through the development of the components industry. Its development can be seen in two distinct phases.[8]

In the first phase, stretching from 1962 to 1967, the Chilean parts industry was dominated by small, adaptable, multi-functional plants, for which components production for the motor vehicle industry constituted no more than another sideline. According to the Instituto de Costos,[9] of a total of 255 Chilean parts producers in 1968, only ten had been formed as a direct result of the demands created by motor vehicle production. Many parts producers were

Table 7.1: Location of Plants Producing Components for the Chilean Motor Vehicle Industry, 1969

Classification of Parts	Santiago	Arica	Other Provinces	Total
Parts from non-ferrous metals	41	7	2	50
Parts from steel and other metals	162	10	5	177
Rubber-based parts	21	—	—	21
Upholsterers	15	3	2	20
Glass producers	3	—	—	3
Plastic parts	21	2	1	24
Others	6	2	—	8
Total	271	25	10	306

Source: Corporación de Fomento de la Producción

by necessity small firms operating in cheap buildings but with highly skilled labour and forceful entrepreneurs. This was due to two factors. First, there was a large number of final assembly firms producing small numbers of cars (at an average of only 430 per firm in 1965). Secondly, each firm produced more than one model (even by 1969, 12 firms were producing 23 models), which meant that the market for each individual part was very small. Thus, there was little economic justification in creating specialised parts producers given the low level of individual parts demand on the one hand, and the large investments that such plants would need, on the other. Instead, the small productive runs could best be accommodated by small, adaptable firms where investment was at a minimum but where skilled labour and entrepreneurs could produce the final part relatively cheaply. This had the locational corollary that such reserves of small-scale entrepreneurs and skilled labour could only be found in the capital, Santiago. By 1969, of the 306 Chilean parts producers, 271 were located in Santiago (Table 7.1). Santiago had a monopoly in the production of rubber and glass parts, and contained the great majority of other parts production. Only 25 small-scale producers had started manufacturing in Arica (mainly small metal parts) and only ten producers were located elsewhere in the country.

The second phase in the development of the parts industry was associated with the growing interest of the international vehicle companies in Chile. As assembly companies were forced to incorporate more nationally produced parts in their assembly, the technological level of domestic parts production rose. The

Table 7.2: Location of Important Vehicle Component Plants in Chile, 1971, According to Industrial Type

Industrial Type	No. of Plants	Location			
		Santiago	Arica	Valparaíso	Rancagua
Plastic parts	8	8	–	–	–
Glass	3	3	–	–	–
Rubber parts	6	6	–	–	–
Steel parts	5	5	–	–	–
Non-ferrous parts	3	3	–	–	–
Electrical parts	11	8	–	2	1
Metal treatment	7	7	–	–	–
Bodywork	13	12	1	–	–
Forgings	5	3	1	1	–
Upholstery	4	2	1	1	–
Total	65	57	3	4	1

Source: CORFO – Guia de fabricantes de piezas y partes para la industria automotriz, 1971

production of these more complex parts came to be beyond the capacities of the small back-street plant. Thus, the international companies financed, helped to finance, or encouraged the development of specialist parts producers — such as Femsaco, the producer of electrical instruments in Rancagua, the Armstrong suspension factory in Santiago, the gearbox factory in Los Andes, and INSA, the nationalised tyre factory in Santiago. Furthermore, the international companies promoted production within their own factories.

Inevitably the net locational result was a further strengthening of the concentration of parts production in Santiago. Of the 311 plants producing components for the Chilean motor vehicle industry in 1971, 281 were located in Santiago, 21 in Arica and only 9 in other areas. Obviously, these 311 establishments included a wide range of plant — from the producer of nuts and bolts for one model to the production of electrical instruments for both national and international markets. But when the Corporación de Fomento de la Producción (CORFO) proceeded to identify the 65 most important plants in Chile, classifying them by type and location, the pattern revealed was even more concentrated (Table 7.2): 57 of the 65 key component plants in Chile were located in Santiago. No major plastics, glass, rubber or metal component plants existed elsewhere. Of the eight important component plants outside

Santiago, only three were located in Arica and these were either closely linked with the international companies (such as Citroën's body work and forgings plants or else were the product of government planning through the national development board, CORFO, — for example, the upholstery plant in Arica, Enatap, was totally owned by CORFO. No major national private firm had decided to locate a components plant in Arica.

Motor vehicle distributors were equally concentrated in Santiago. By the end of the 1960s, 70 per cent of all vehicle sales in Chile were carried out by distributors in Santiago. A situation in which 30–50 per cent of the FOB value of a car was transported the 2,000 kilometres from Santiago to Arica only to be followed by 70 per cent of cars being taken back to the capital, brought home the notable lack of success of the motor vehicle industry as a basis for growth pole development in northern Chile. The industry's supposedly advantageous characteristics of large size, purchasing dominance and high levels of interaction had failed to generate significant local multipliers.

The multinational vehicle companies and their subsidiaries soon reacted to the failure of Arica as a motor vehicle growth pole. In the late 1960s, three vehicle companies sold their plants to incoming electronic industries and built new plants near to Santiago. Fiat built a plant at Rancagua, Ford at Casablanca, and the Chilean company holding the Peugeot license started producing at Los Andes (see Figure 7.1).

This lack of local multipliers from a decentralised motor vehicle industry has also been witnessed in Trujillo, Barcelona, Cordoba and Belo Horizonte. In each case, the small firms and specialist component producers of the primate city have been the major beneficiaries of the supply and service linkages created by the establishment of an assembly plant in a decentralised location. In Fiat's Betim plant near Belo Horizonte, 84 per cent of parts suppliers and 65 per cent of the value of the parts come from the São Paulo agglomerations (see Table 7.3). The majority of Minas Gerais's 17 per cent contribution to the value of parts comes from the Fiat foundry, built adjacent to the assembly plant itself. In the case of Cordoba, a comparable 'flight' of motor vehicle production to that from Arica has taken place.

In 1959, Cordoba accounted for 86 per cent of Argentine vehicle production.[10] However, with Fiat later switching production to a new Buenos Aires plant, and General Motors and Ford following

Table 7.3: Location of Suppliers to Fiat's Plant in Belo Horizonte, Minas Gerais

Location of Supplier	Suppliers		Parts		Value of Parts
	No.	%	No.	%	%
São Paulo	489	84.4	3,696	74.4	64.7
Minas Gerais	49	8.5	784	15.8	16.9
Rio de Janeiro	29	5.0	430	8.7	2.7
Italy	5	0.9	32	0.6	12.6
Other Locations	7	1.2	26	0.5	3.1
Total	579	100.0	4,968	100.0	100.0

Source: A. Segre, The Location of the Automobile Industry in Latin America. The cases of Belo Horizonte and Cordoba. IGU Commission on Industrial Systems Conference, São Paulo, 1982.

this locational preference, Cordoba soon finished being the vehicle town of Argentina; by 1971 it accounted for only 18 per cent of Argentine production — mainly the Renault plant that remained. Attempts at industrial decentralisation through the use of the motor vehicle industry have therefore met with little success in Latin America. Even the largest and most successful decentralised plant, that of Fiat at Betim, with an annual capacity in excess of 200,000 vehicles a year, suffers from the extra transport costs incurred in bringing 3,696 different items from parts producers in São Paulo.

Decentralisation of Other Industries

It could be argued, however, that the motor vehicle industry is not a good selection as a decentralised industry *if* multipliers are not concentrated in the local region. This is because vehicles are bulky items for which transport costs are significant both in transferring parts from component plants to assembly plant and in transferring the finished vehicle to the market. It has been estimated that the increased costs involved in shipping parts from Santiago to Arica and the finished vehicle from Arica to Santiago augmented the final price of the finished vehicle in Chile by an average of 10 per cent.[11]

If the predictions of growth pole theory are rejected and if it is accepted that most of the multipliers of a plant in a peripheral region will be felt outside of that region, a more appropriate candidate for decentralisation would be an industry in which total transport costs were a much smaller proportion of total costs. High technology industry such as the assembly of electronic consumer goods would fulfil these requirements. In the case of the television industry of Arica, one survey found that the extra transfer costs incurred by the Arican plant in comparison with a similar plant in Santiago was equivalent to only 2.1 per cent of the total cost of a television in 1969.[12] Such an overcost would not appear excessive for Chilean consumers purchasing electrical goods within a framework of import substitution. Meanwhile, the television industry became the major industry of Arica in the early 1970s, employing 3,000 workers at its peak in 1972. From the taxes on the increased commercial and industrial activity caused by the television industry, the regional development body of Arica, the JAA, was able to finance considerable investment in both infrastructure and social development.

The Resource Growth Pole

From the mid-nineteenth to the mid-twentieth century, Latin America's dominant role in the world economy was as the provider of mineral and agricultural resources for the industrialised countries of the core. The functions of the resource location within Latin American countries were normally restricted to primary activities. In mining, this consisted of the extraction of the mineral and some basic processing and refining if the mineral content of the ore was low. In agriculture, more secondary activity resulted, but rarely was this the basis for further, more diversified industrialisation. Resource locations in Latin America, therefore, were rarely able to build up an industrial complex based around their material attractions. In such a way, the location of industry in Latin America was radically different from that in Europe where the coal and iron ore field proved very attractive to a wide range of industry.

One major factor behind the lack of industrialisation at resource locations in Latin America was the foreign ownership of resources. Foreign firms preferred to process, refine and combine resources at locations within the firm's country of origin or near to substantial markets. However, with the widespread nationalisation of such foreign firms in Latin America and the creation of state firms to control the production and marketing of resources, different attitudes have developed towards the resources location. In contrast to previous trends, it has become the policy of government to concentrate on localising as many related processes as possible at the resource location. For example, whereas once iron ore would have been simply exported, governments now attempt to create steel plants and steel-using industries near the iron ore location. Such a policy has obvious links with growth pole theory. Instead of a growth firm providing the impetus to the location of related activities, the resource and government commitment to invest in the further processing of that resource theoretically supply the initiative for related private investment.

A major constraint on the expansion of manufacturing activity at resource locations has been government finance. This has been particularly noticeable if activities related to the resource have been previously located elsewhere in the country. Nevertheless, one notable resource growth pole has been financed by one of the two OPEC countries in Latin America, Venezuela. The resource growth pole is that of Ciudad Guayana, at the junction of the Caroni and Orinoco rivers in Eastern Venezuela. In 1960, the recently elected President Betancourt created the Corporación Venezolana de Guayana (CVG) in order to develop the resources and promote the industrial development of this peripheral region.

The region does have extraordinary resources. There are rich deposits of high grade iron ore and promising possibilities for the mining of manganese, nickel, chromium, gold, industrial diamonds, bauxite and aluminium laterite. Within sixty miles of Ciudad Guayana there are fields of petroleum and natural gas. The settlement is on the banks of the Orinoco river which provides direct access to the ocean. Running through the heart of the city is the Caroni river, which has a hydroelectric potential of about 10 million kilowatts. Furthermore, the area presents considerable opportunities for the development of agriculture and the exploitation of forestry products.

The extraction and processing of these resources is now almost exclusively controlled by Venezuelan state firms. Iron ore and petroleum mining were nationalised in 1975 and 1976 respectively. Steel production has been organised by the state ever since 1962 when the first steel plant went on stream in Ciudad Guayana. In 1964 the Siderurgica del Orinoco (SIDOR), the state steel firm was formed as a subsidiary of the CVG. The original SIDOR plant, with a 0.7 million ton capacity, was designed to produce entirely non-flat and tubular products, particularly bars and rods for the construction industry and seamless pipes for the oil industry.[13] However, the main growth area in the 1960s was in flat products, owing to the development of metal-using consumer-durable industries in the central Caracas-Valencia axis. The position was rectified in the early 1970s when a flat-products mill was added.

However, by 1974, with Venezuelan consumption reaching two million tons per annum, and Venezuelan production totalling one million tons, major imports of steel were again having to be made. Venezuela was returning to a position of exporting large amounts of iron ore and importing sizeable quantities of steel. The trade was not beneficial for the country. Venezuela's exports of over 25 million tons of iron ore did not earn enough to pay for its imports of one million tons of steel in 1974.[14]

As a result an ambitious National Steel Plan was approved in 1974. Its cornerstone was the investment of US $3,600 million in a steel complex in Ciudad Guayana, where annual steel capacity was planned to increase to 5.5 million tons in the early 1980s. The principal installations of SIDOR's Fourth Plan were a 6.6 million tons p.a. pelletising plant (for the treatment of local iron ore from Cerro Bolivar), eight direct-reduction modules with a combined capacity of 5.4 million tons of sponge iron a year, and six 200-ton and four 150-ton electric ore furnaces coupled to continuous casting machines for the production of slabs and billets.[15] The reason for such a complex technological process was that the direct-reduction modules could be fuelled by local natural gas and the electric ore furnaces by the cheap hydroelectricity of the planned Guri dam. As all three inputs (iron ore, natural gas, hydroelectric power) were both locally abundant and could be priced and controlled by firms in the state sector, the finished steel should likewise be cheap and benefit from considerable comparative advantage in world trade.

The Ciudad Guayana growth pole came to be a cornerstone of Venezuela's Fifth Plan (1976–80). The plan was distinguished by the decision to invest the huge income reserves of Venezuela, a result of the 1973–4 quadrupling of oil prices, in industry and infrastructure. It was the first time the words of Venezuela's most influential oil minister, Perez Alfonso, 'sembrar el petroleo' (sow the oil) were actively being heeded. Of a total budget of $30 billion, $23 billion were destined for industry and infrastructure; one third of this was channelled into the Ciudad Guayana project. After the steel plant, the next largest investment was planned for hydro-electricity. The Guri project, which involves a raising of the elevation and a widening of an existing dam, had an initial budget of $4,000 million and should have a capacity of 9,000 megawatts in 1985. This will not only provide electricity for the SIDOR project but also for the expanding aluminium complex (see Chapter Four).

It is still too early to judge the industrial success of Ciudad Guayana, the most ambitious resource growth pole outside the Soviet Union. SIDOR has had major technological and financial problems in its start-up phase and has still not significantly increased its annual production level above one million tons per annum. As a result, imports of steel are still high and there is little locational advantage for steel-users to locate in Ciudad Guayana. One of SIDOR's major medium-term problems is the lack of electricity as the Guri project has fallen disastrously behind schedule. According to the Venezuelan Engineers Association, Venezuela could have an electricity deficit as high as 3,500 megawatts in 1985.[16] Most of this deficit will be spatially concentrated in the electricity-intensive Ciudad Guayana industrial complex. Furthermore, unemployment is beginning to increase rapidly, as the basic construction phase finishes (only Guri's construction remains) and as industrial production and employment (apart from in the aluminium sector) fail to expand significantly.

Nevertheless, it seems likely that Ciudad Guayana will not attract a herbaceous border of metal-using industries to its large, capital-intensive plant. The largest national metal-using group, SIVENSA, prefers to expand near the market in the Valencia-Caracas axis rather than in Ciudad Guayana. The government has only been able to pressurise multinational enterprise to locate plants in the region. The best example of this is the emerging heavy vehicle complex in Ciudad Bolivar where Fiat diesel engine and medium truck plants are planned to join Mack's diesel engine and John Deere's tractor plants.[17] Meanwhile, the Venezuelan governments of the 1980s, after witnessing so much government investment in the region in the 1970s, are cutting back on planned investment in related and new activities. Only investment in the large, unfinished projects is being maintained. With government investment consolidating and national enterprise prevaricating, it seems probable that Ciudad Guayana will not significantly change the existing industrial concentration of production and employment that occurs in Venezuela.

Industrial Decentralisation in Brazil and Mexico: a Comparison

Brazil and Mexico are the foremost industrial countries of Latin America. They have advanced the furthest along the path of import substitution and have highly diversified industrial structures as a result. Some writers have argued that their policies towards the institutional development of industry have been very similar.[18] Furthermore, they have remarkably similar patterns of spatial concentration of industry. Both countries have approximately 50 per cent of their manufacturing production and employment generated in one, sprawling, multi-centred metropolitan area — functionally interlinked and spatially continuous. The major difference is that whereas Mexico City is a federal capital, São Paulo is not.

Furthermore, both countries are likely to generate an even more spatially concentrated industry in the future, given that the fast-growing, dynamic industries are more concentrated in the major city than are the more traditional slow-growth sectors. In Brazil, 75 per cent of employment in the booming electrical and vehicles sectors and 70 per cent in the plastics and rubber industries corresponded in 1970 to São Paulo state; in the same year, only 13 per cent of employment in the wood industry and 31 per cent in the massive food industry was generated from São Paulo state.[19] Meanwhile, in Mexico, 56 per cent of all new import-substituting industry was located in the major metropolitan area in the early 1960s.[20]

Both countries are also characterised by a well-established decentralised location for industry, based fundamentally around the iron and steel industry. Monterrey in NE Mexico has no local resources, but according to Kennelly,[21] constitutes a classic least-transport-cost location between a variety of regional and extra-regional resources and markets — iron ore from Durango, coke from Sabinas, gas from Texas, oil from Reynosa and Tampico, and markets divided between Monterrey and Mexico City. Balán further stresses the importance that the US market has occasionally had in the growth of Monterrey's metallurgical industry.[22] Monterrey's metllurgical industry has been based on one large plant, the Fundidora but Belo Horizonte's growth has been linked to three plants, only one of which, the Mannesmann plant, is local. The other two plants, those of Belga Mineira at Monlevade and USIMINAS at Ipatinga, are situated to the east of Belo Horizonte on the River Doce, near to the largest iron ore deposits in Brazil.[23]

These two industrial cities of Monterrey and Belo Horizonte also share a belief in the entrepreneurial capabilities of their citizens — called the 'regiomontanos' and 'mineiros' respectively. This is particularly the case in Monterrey where the 'regiomontanos' are 'perceived as an industrious people' with entrepreneurs 'that have continually reinvested a large part of the profits from their operations'.[24]

How have schemes of industrial decentralisation fared in these two countries where industrial centralisation is the norm and decentralisation the exception? There has certainly been a difference of approach. Brazil has concentrated decentralising efforts in a well-funded scheme in the Northeast, while Mexico has developed many smaller schemes in different regions and different sectors.

The Brazilian Northeast, with approximately one-third of the country's population, is the poorest region of Brazil. Its problems have been the preoccupation of governments for over a century, particularly after the occasional droughts that the region has suffered. After the 1958 drought, the government created an integrated regional development agency, the Superintendencia do Desenvolvimiento do Nordeste (SUDENE). The creation of SUDENE marked a major change in policy attitudes towards the Northeast. Previous government initiatives had been confined to the agricultural problems but SUDENE's objectives included that of promoting industrialisation. It was anticipated that industrial growth would both diversify the regional economy and absorb labour.[25] Capital for industrial growth was provided by tax credit schemes deriving from Articles 34 (1961) and 18 (1963).[26] Under this scheme, Brazilian corporations could reduce their tax liabilities by up to 50 per cent if they invested this amount in industrial projects in the Northeast. SUDENE had the role of reviewing projects, but if an industrialist met all the SUDENE criteria, he could obtain up to 75 per cent of the capital he required from 34/18 funds. Other capital available from the Banco do Nordeste and from individual states and 'municipios' anxious to promote industrialisation within their boundaries meant that up to 87.5 per cent of proposed capital investment could be obtained from public sources. The possibility of such relatively large capital inputs being derived externally for the construction of new or modernised plants attracted large numbers of industrialists, particularly in the late 1960s when Brazil experienced high economic growth rates. By the end of 1970 some 750 industrial projects had been approved.

Nearly two thirds of new investment approved by SUDENE took place in the chemical, metallurgical, electrical, machinery and vehicle sectors.[27] Nevertheless, three major problems have been perceived that cast doubt on the overall success of the scheme.

First, there has been a marked spatial concentration of industrial plants in the three major cities of the Northeast — Recife, Salvador and Fortaleza. According to Dickenson, 41 per cent of the projects approved by 1973 were located in these three cities, the majority destined for the industrial estates specially constructed in Recife (Cabo, Paulista), Salvador (Aratu) and Fortaleza to attract southern entrepreneurs.[28] Meanwhile, only 3 per cent of new investment was channelled into the whole of the two poorest states of Maranhão and Piauí.

Secondly, the new industrial economy of the Northeast is emerging as an appendage of that of Brazil's core area. In the metallurgical, electrical, machinery and vehicle sectors, more than 80 per cent of domestic inputs come from the Centre-South region.[29] The significant backward linkages that are characteristic of these sectors (e.g. increased demand for components and machinery) therefore benefit the Centre-South region and not the Northeast. The chemical sector is more integrated into the Northeast economy, linking in with the oil fields, the Mataripe oil refinery near Salvador and a local petrochemical growth pole that was designated as such in 1971. One survey showed that over 50 per cent of domestic inputs into the Northeast chemical sector came from that region.[30]

For the metallurgical industries, however, the situation is exacerbated by the fact that the majority of firms in these sectors sell 50 per cent or more of their output outside of the Northeast, mainly to the Centre-South. Such a pattern can be attributed to the level of demand in the Northeast, due to its poverty rather than lack of population, and the creation of capital-intensive firms. Because the Tax Credit Scheme of Articles 34/18 made capital abundant and cheap, most industrialists maximise capital rather than labour inputs in the creation of new plants. As capital-intensive plants have high fixed costs, low cost production only results from long production runs. This need for long production runs normally means that firms have to rely on much larger markets than the Northeast can offer and hence the prevalence of finished goods being sent to the major markets of the Centre-South. Thus, with the majority of inputs coming from the Centre-South and the major portion of finished products returning there, the metallurgical sector of the Northeast's 'new' industrial economy is very much an appendage of the Centre-South. The chemical sector is better integrated into the regional economy in terms of inputs, but the major markets for the sector's products are nevertheless in the Centre-South.

The third and major problem of the 34/18 scheme has been its inability to meet one of the two aims of SUDENE's industrial policy, that of absorbing labour. Between 1960 and 1970 there was a net increase of only 55,000 in the size of the Northeast's industrial labour force.[31] The easy acquisition, abundance and cheapness of capital meant that the new capital-intensive plants generated little new employment while the modernisation of older plants, particularly in the textile sector, actually reduced employment. Between 1960 and 1970, the four traditional industries of the Northeast, textiles, leather, tobacco and soap, experienced a net loss of 23,500 workers, due to the radical modernisation of the traditional plants.[32] It is interesting to note that the major increase in absolute numbers of jobs did not come in the metallurgical and chemical sectors but in sectors more closely geared to the regional economy. Thus, 33,000 new jobs were created in the food-processing and non-metallic mineral sectors, the latter sector providing the majority of the inputs for the construction industry.

Despite being well-funded by a tax credit scheme, the programme of industrial decentralisation in the Northeast has been unable to counteract significantly the powerful centralising forces that govern the location of industry in Brazil. Many of the new industrial plants of the Northeast are assembly branch plants of firms based in the Centre-South; the industrial economy of the Centre-South therefore receives the majority of backward linkages and the resultant multipliers. Meanwhile, the Northeast's share of Brazilian industrial production continued to fall during the 1960s from 6.9 per cent in 1959[33] to 6.4 per cent in 1970.[34] The net employment increase of 55,000 in ten years had little impact in denting the high rates of underemployment in a regional population of thirty million where 69 per cent of the urban employed receive the equivalent of one minimum salary or less.[35]

Mexican regional planners have not had to deal with such a major problem as the Brazilian Northeast and attempts at industrial decentralisation have been restricted to a series of *ad hoc* measures such as the creation of industrial estates (Torreón) and new industrial cities (Monclova). There has been little attempt to decentralise industry away from the Mesa Central. In the south, the river basin projects have had little impact in attracting manufacturing industry — indeed their share of Mexican manufacturing production actually declined in the 1960s.[36] On the northern border, the border industrialisation programme has attempted to attract assembly plants of United States companies to take advantage of the large labour-cost differential between Mexican and US workers (see Figure 7.2). According to Baerresen's study, rates in northern Mexico are between one third and one fifth those in the United States.[37] As a result, and to some advantage for Mexico, labour-intensive industry is characteristic of the fourteen border towns that in 1971 contained 215 industrial plants. The major types of labour-intensive assembly industries were the production of parts, such as semiconductors and memories, for the electronic, office machine and television industries (39 per cent of production), the manufacture of toys and dolls (12 per cent), clothing (12 per cent), and the production of televisions (9 per cent) and scientific instruments (4 per cent).[38] However, the dominance of repetitive assembly operations in the industrialisation of the northern border towns has produced an unbalanced labour market, with approximately 85 per cent of manufacturing workers being women and with male unemployment nearly doubling as a result.[39] Such a phenomenon has done little to stop the flow of illegal male migrants to the United States, one of the reasons why the United States was willing to waive most customs duties on the import of goods from the border industrialisation programme.[40] Meanwhile, the assembly plants are closely geared to US industry and inputs. Indeed, US customs only waives duties on the assembly of US inputs; the customs duties on other inputs have to be negotiated or charged at the full rate. The Mexican border's industrial economy is therefore distinctly artificial, linked strongly to the plants and firms of a neighbouring country and generating few local multipliers. Its long-term future is in the hands of international arrangements over which it has little influence.

The experience of industrial decentralisation in Brazil and Mexico has been very different. However, in both cases, the programmes of industrial decentralisation have not provided a basis for circular and cumulative processes of regional economic growth. Most of the significant multipliers generated by such schemes have been felt outside the region being assisted.

Figure 7.2: The Mexican Border Industrialisation Programme

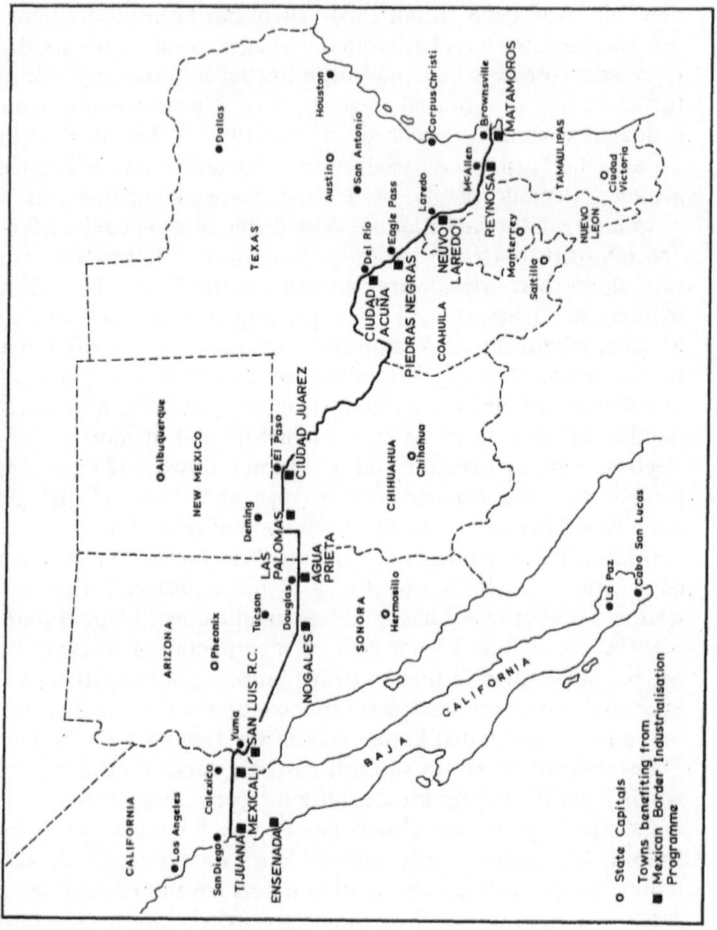

Existing Processes of Industrial Decentralisation

The large primate city does however have some disadvantages for the location of industry in Latin America. The cost of land is normally higher and tends to dissuade land-intensive industries. The increasing congestion of many large Latin American cities makes transportation more difficult both for people and products. The costs of housing, health, recreation and labour *can* also be higher in the major city.

The potentially high costs of locating or even maintaining an industry in the primate city can be found in the Venezuelan capital, Caracas. In 1976, the Caracas agglomeration had a population of 2.6 million, crowded into an elongated but narrow basin whose east-west axis measures 10 miles from Catia to Petare but whose north-south dimensions rarely exceeded two miles. To the north, the steep Cordellera de la Costa prevents any expansion, and indeed the area has been designated as a national park. To the east and west, high land similarly excludes development. It is only in the south where five valleys penetrate hills of more moderate but still significant elevation that outward expansion of the city can still occur — an area that because of high costs of construction, land and transport is becoming an exclusive residential zone for the middle classes.

The unavailability of flat land for industry after 1966, the concomitant high land prices and congestion have acted as powerful decentralising forces for Venezuelan manufacturing. In no sector is this better represented than in the motor vehicle assembly industry where the use of large areas of cheap flat land is at a premium. As Figure 7.3 demonstrates, nine vehicle assembly plants are now located outside Caracas in the axial belt stretching from Las Tejerias to Valencia and Morón. Only the original General Motors plant remains in the Caracas agglomeration, in the southern industrial suburb of Antimano.

The movement of assembly plants into the axial belt effectively began with the introduction of Ford as a new producer in 1963 and that company's decision to locate its new plant at a green field site on the outskirts of Valencia. Within four years Ford had become the major vehicle manufacturer of Venezuela, overtaking the two major US producers, Chrysler and General Motors, in the process. Chrysler and General Motors found it difficult to react to Ford's increases in market share, partly because of the restrictions their plant sites put on productive expansion. This was felt most acutely by Chrysler which had a site near central Caracas, in which further expansion was impossible. As a result, Chrysler decided to sell its plant in Caracas, followed Ford's locational decision and located a new plant on a site in Valencia with ample room for expansion. The selling of the Caracas plant almost covered the costs of the land and new plant in Valencia.[41] As a result, the production of Chrysler vehicles was able to expand and by 1973 Chrysler had pushed General Motors into third place in terms of production. This position lasted until 1979 when General Motors bought out the Chrysler plant and started producing its own vehicles there.[42] Since then, General Motors has been concentrating production at the ex-Chrysler plant in Valencia and using the Antimano plant for more specialised activities.

Industrial decentralisation has therefore been occurring in Venezuela but in the form of movement along the Caracas-Valencia axial belt. The well-established towns of Valencia and Maracay have been the main recipients of new and relocating industrial plants. But small towns, such as La Victoria, and even large villages, such as Mariara, have attracted significant industrial enterprises. Meanwhile government planners have been intent on promoting the short-distance movement of industry out of Caracas to a collection of new towns in the middle Tuy valley and a large planned town to the east, Ciudad Fajardo, that will be a bi-nodal agglomeration combining the settlements of Guarenas and Guatire (see Figure 7.3). Nevertheless, in 1974, 47 per cent of Venezuela's 300,000 manufacturing workers were employed in the Caracas agglomeration (including the port of La Guaira).[43] Increasing numbers were working in flatted factories such as those in La Urbina, an industrial estate built in the far east of the Caracas basin near Petare. Meanwhile, only 90,000 manufacturing workers (30 per cent of the total) were employed in the other settlements of Venezuela's axial belt in 1974.[44]

This process of industrial decentralisation from Caracas into the axial belt is being encouraged further outwards by the Industrial Decentralisation Policy of 1974.[45] Few benefits can be achieved by any but high priority industry in Caracas (Area A) or the Caracas-Valencia-Puerto Cabello axis (Area B) (see Figure 7.4). Benefits in the form of favourable credits and profit tax reductions can be acquired for most industrial plants locating in Marginal Area B (non-industrial areas to the south of the Puerto Cabello-Valencia-

Figure 7.3: The Venezuelan Axial Belt

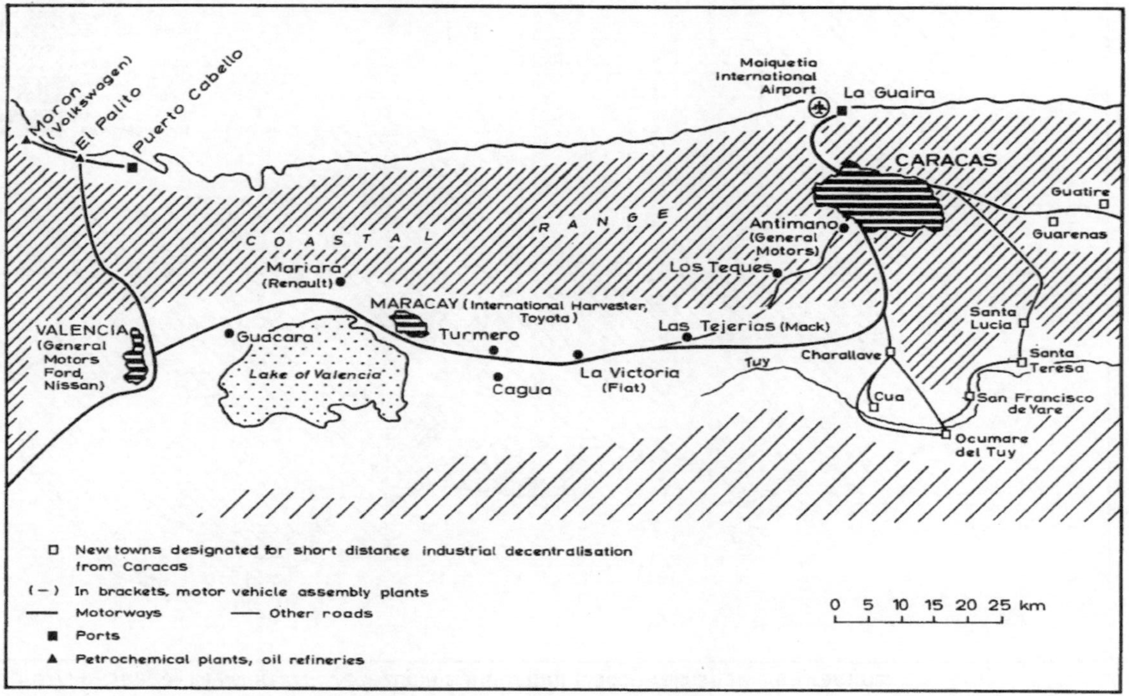

Figure 7.4: Spatial Divisions of Venezuela's Industrial Decentralisation Programme

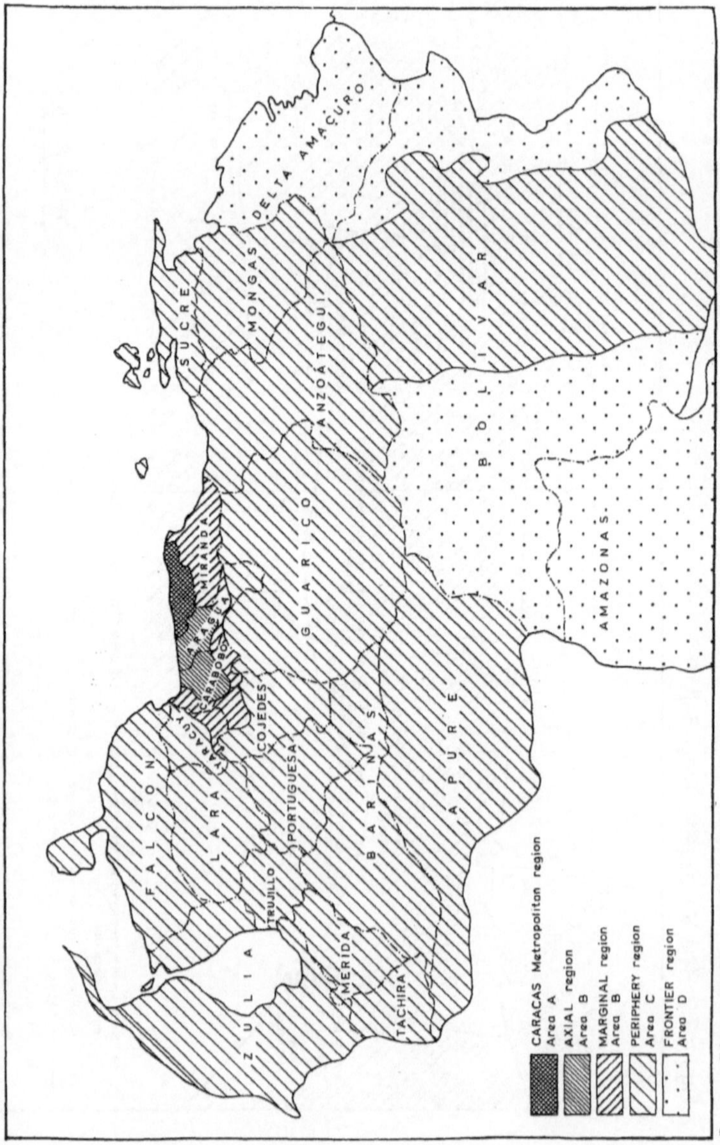

Source: Dirección General Sectorial de Industrias, *Dirección de Promoción y Desconcentración Industrial* (Caracas, 1977)

Caracas-Guatire axis) and Areas C and D. The basic idea is to extend the industrial belt both eastwards (to include Barcelona and Cumaná) and westwards (to include Barquisimeto and San Felipe). Industries whose development is closely controlled by the state will assist in the process. For example, in the motor vehicle industry, General Motors' new engine plant will be built at Barquisimeto and a Fiat engine plant at Barcelona, while Pegaso's new truck plant is already producing at Cumaná.

Short-distance industrial decentralisation from the primate city and government encouragement of the phenomenon are common to many other Latin American countries. In Mexico, short-distance industrial movement has been prevalent along the Mexico City-Puebla axis. At the same time, Mexican governments have encouraged short-distance decentralisation by establishing large industrial estates at Queretaro and Irapuato, by creating new industrial cities at Ciudad Sahagún and Cuernavaca and by planning an industrial zone of the west, south of Guadalajara, and including the towns of Octlán and La Barca.[46] As a result most of the effective industrial decentralisation of Mexico has been confined to the Mesa Central where 70 per cent of Mexican industrial production is generated.[47]

The twentieth century development of Latin America's major industrial agglomeration, São Paulo, has been characterised by the steady onion-like growth of industrial zones around the original centre. The 'municipio' of São Paulo still represents the largest concentration of industry in Brazil with 24 per cent of all industrial workers in Brazil and over half of all employment in the electrical and plastics industries. Surrounding and including the municipio is Greater Sao Paulo, reflecting industrial decentralisation in the 1950s and 1960s when the ABC town of Santo André, São Bernardo do Campo and São Caetano do Sul became attractive for a wide range of industry but particularly that of motor vehicles. The vehicle plants of Chrysler, Ford, Mercedes-Benz, Saab-Scania, Toyota and Volkswagen are concentrated in Brazil's vehicle capital of São Bernardo with the General Motors plant in neighbouring São Caetano.[49] In the 1960s and 1970s, however, new industries have located in and older industries moved out to the smaller inland towns of São Paulo state. According to Boisier, it is these towns of between 200,000 and 500,000 inhabitants where industrial productivity is highest in terms of urban size.[50] São Paulo's industrial engine has moved out along three axes (see Figure 7.5): to the

Figure 7.5: The Industrial Core of Brazil

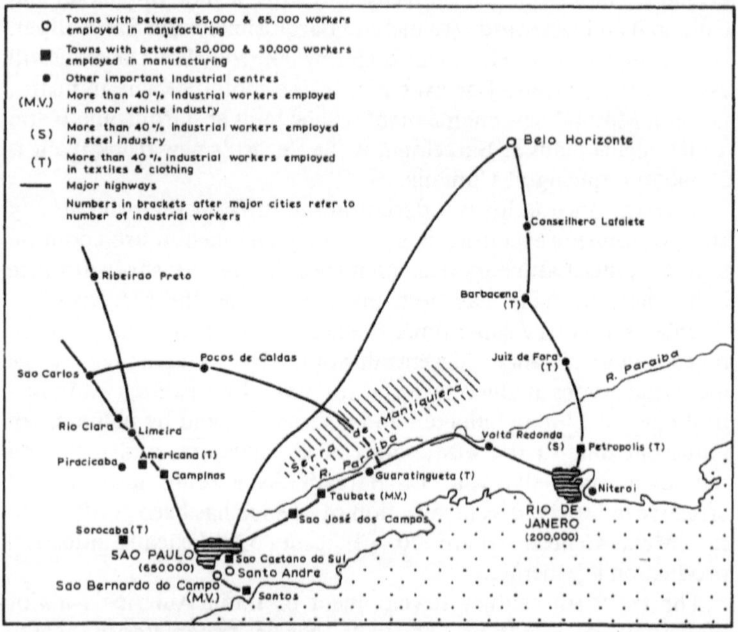

northeast along the Paraíba valley to incorporate the towns of São José dos Campos and Taubaté, where in particular new component and vehicle assembly plants are locating; to the northeast along the Paulista railway line to incorporate such towns as Campinas, Piracicaba and Americana, where engineering and textiles stand out in a wide range of industry; and to the west to incorporate such towns as Sorocaba, where non-metallic minerals, food processing and textiles are the major sectors.[51]

Such short-distance decentralisation has become most associated with the larger more dynamic industrial countries of Brazil and Mexico, where import substitution industrialisation has progressed furthest and has created a highly diversified, closely interlinked urban-industrial structure. Within a national framework of rapid economic growth, both multinational and domestic industrial firms continually aim to expand production and diversify into new often related areas. Successful multinational and domestic firms can therefore become involved in a large number of locational decisions, such as those of plant expansion, new plant creation and relocation of old plant. Many of these locational decisions will take place within the spatial framework of the primate city and its environs as this is not only where the majority of industrial firms are based but also where the fastest growing are developing. In this sense, short-distance decentralisation can best be envisaged as an intensification of the process of industrial centralisation and as an expansion of the industrial core out from the boundaries of the primate city to a wider functional region beyond.

Industrial decentralisation has been a declared objective of governments and planners since the 1960s. The initiation of Brasilia, SUDENE, Ciudad Guayana, the Arica free port and Mexican river basin developments can be traced to this decade. Yet during the last thirty years the process of spatial concentration of industry in the primate city of Latin American countries has intensified markedly. Government-financed schemes have had limited and sometimes negative effects, even when based on an impressive regional resource base. Growth pole planning has invariably had a greater impact on the central city than on the economy of the region it was supposed to assist. The major decentralising phenomenon, that of short-distance decentralisation from the primate city, can best be seen not as a successful form of industrial deconcentration but as a further strengthening and intensifying of the process of centralisation.

References

1. R. N. Gwynne, 1978, *Industrial Decentralisation in Chile*. (Unpublished doctoral dissertation, University of Liverpool)

2. A. Gilbert, 1976, The arguments for very large cities reconsidered, *Urban Studies*, 13, pp. 27–34; H. W. Richardson, 1976, The argument for very large cities reconsidered: a comment. *Urban Studies*, 13, pp. 307–10; A. Gilbert, 1977, The arguments for very large cities reconsidered: a reply. *Urban Studies*, 14, pp. 225–7

3. F. Perroux, 1950, Economic space: theory and applications, *Quarterly Journal of Economics*, p. 64

4. F. Perroux, 1971, Note on the concept of growth poles, in I. Livingstone (ed.), *Economic policy for development: selected readings*. (Penguin, Harmondsworth)

5. J. R. Boudeville, 1966, *Problems of Regional Economic Planning*. (Edinburgh University Press, Edinburgh)

6. A. O. Hirschman, 1958, *The Strategy of Economic Development*. (University Press, New Haven, Conn.)

7. R. N. Gwynne, 1978, Government Planning and the Location of the Motor Vehicle Industry in Chile, *Tijdschrift voor Econ. Soc. Geog.*, 69, pp. 130–40

8. R. N. Gwynne, Industrial Development in the Periphery

9. Instituto de Costos, 1969, *Estudio sobre la industria automotriz*. (Santiago)

10. A. Segre, 1982, The Location of the Automobile Industry in Latin America. The cases of Belo Horizonte and Cordoba, IGU Commission on Industrial Systems Conference, São Paulo
11. R. N. Gwynne, 1980, *Import Substitution and the Decentralisation of Industry in Less Developed Countries: The Television Industry in Chile, 1962–1974*. (Department of Geography, University of Birmingham, Occasional Publication No. 12)
12. Ibid.
13. P. J. West, 1979, Venezuela: the iron and steel industry, *Bank of London and South America Review*, 13, 3, pp. 138–48
14. The Economist, 1975, The land of material interests — a survey of Venezuela, December 27th
15. P. J. West, Venezuela: the iron and steel industry
16. Latin America Regional Report, The Andean Group, 16/5/80
17. R. N. Gwynne, 1979, The Venezuelan automobile industry, *Business Venezuela*, 64, pp. 24–38
18. G. Gereffi and P. Evans, 1981, Transnational Corporations, Dependent Development and State Policy in the Semiperiphery: A Comparison of Brazil and Mexico, *Latin American Research Review*, 16(3), pp. 31–64
19. J. Dickenson, *Brazil*
20. A. M. Lavell, 1972, Regional Industrialization in Mexico: Some Policy Considerations, *Regional Studies*, 6, pp. 343–62
21. R. A. Kennelly, The Location of the Mexican Steel Industry
22. J. Balan, H. L. Browning and E. Jelin, 1973, *Men in a Developing Society: Geographic and Social Mobility in Monterrey, Mexico*. (University of Texas Press, Austin)
23. J. Dickenson, *Brazil*,
24. J. Balán et al., *Men in a Developing Society*
25. A. O. Hirschman, 1968, Industrial Development in the Brazilian Northeast and the Tax Credit Scheme of Article 34/18, *Journal of Development Studies*, 5, pp. 1–28
26. D. E. Goodman, 1972, Industrial Development in the Brazilian North-east: An Interim Assessment of the Tax Credit Scheme of Article 34/18, in R. J. A. Roett, (ed.), *Brazil in the Sixties*. (Vanderbilt Press, Nashville)
27. Ibid.
28. J. Dickenson, *Brazil*
29. D. E. Goodman, Industrial Development in the Brazilian North-east, p. 245
30. Ibid.
31. J. Dickenson, *Brazil*
32. Ibid.
33. Ibid.
34. W. G. Tyler, 1981, *The Brazilian Industrial Economy*. (Heath, Lexington)
35. D. E. Goodman, 1976, The Brazilian Economic "Miracle" and Regional Policy: Some Evidence from the Urdan Northeast, *Journal of Latin American Studies*, 8(1), pp. 1–27
36. D. Barkin and T. King, 1970, *Regional Economic Development: The River Basin Approach in Mexico*. (Cambridge University Press, Cambridge)
37. D. Baerresen, 1971, *The Border Industrialization Program of Mexico*. (Heath, Lexington)
38. Ibid.
39. M. P. Fernandez, 1981, The U.S. — Mexico Border: Recent Publications and the State of Current Research, *Latin American Research Review*, 16(3), pp. 250–67
40. D. Baerresen, *The Border Industrialisation Program of Mexico*

41. Interview with general manager of Chrysler Venezuela, August 1979

42. R. N. Gwynne, The Venezuelan automobile industry

43. Dirección General de Estadisticas y Censos Nacionales, 1976, *IV Encuesta Industrial 1974: Resultados Regionales*. (Ministerio de Fomento, Caracas)

44. Ibid.

45. Dirección General Sectorial de Industrias, 1977, *Politica de Desconcentración Industrial* (Ministerio de Fomento, Caracas)

46. A. M. Lavell, Regional Industrialization in Mexico

47. P. R. Odell and D. A. Preston, 1973, *Economies and Societies in Latin America: A Geographical Interpretation*. (John Wiley, London)

48. J. Dickenson, *Brazil*

49. Ibid.

50. S. Boisier, 1974, Localizacion, tamaño urbano y productividad industrial: un caso de estudio de Brazil, *Revista Latinoamericana de Estudios Urbano Regionales*, 3(9), pp. 57–78

51. J. Dickenson, *Brazil*

PART THREE

INDUSTRIALISATION AND URBANISATION

8 HISTORICAL PERSPECTIVES ON URBANISATION

In 1979 the Population Reference Bureau calculated that the number of people living in urban areas in the Latin American world region (including the Caribbean) was equivalent to 61 per cent of the total population. This figure, however, hides significant regional variations. In temperate South America, urban population constituted as much as 80 per cent of the total, although the tropical and central American regions recorded levels of 60 per cent and 58 per cent respectively. As Table 8.1 demonstrates, the Latin American world region is distinguished from other less developed

Table 8.1: Two Indices of the Percentage of the Total Population of Each World Region that is Urban

	% According to each country's classification (1979)	% People living in places of 20,000 or more (1975)
World Total	39	30
More Developed Areas	68	52
Europe	68	48
N. America	74	65
Oceania	71	57
Soviet Union	62	46
Less Developed Areas	28	22
Africa	25	18
Asia	27	21
Latin America	61	41

Sources: Population Reference Bureau, *1979 World Population Data Sheet* Washington; W. P. Frisbie, *The scale and growth of world urbanization* (Population Research Center, University of Texas, Austin, 1976)

areas by its high levels of urbanisation, levels which are approaching those of many developed world areas. This phenomenon is, however, relatively recent — in 1940, for example, only 19.6 per cent of Latin American populations lived in urban places of over 20,000 inhabitants.

In this third part of the book we will investigate the contribution of industrialisation to this rapid process of urbanisation. However, it is first appropriate to put the modern relationship between industrialisation and urbanisation into some form of historical perspective. The foundations of urbanisation in Latin America go back to pre-colonial times and colonial urban development dates from the early sixteenth century. The following chapter examines the colonial and nineteenth century heritage of modern urbanisation in Latin America.

The Colonial Legacy

In contrast to the gradual evolution of the North American urban system over three centuries, the Latin American urban system was largely created in less than a century. Indeed what were to become some of the principal urban foci of Spanish Latin America were founded between 1520 and 1541 — Mexico City, Puebla, Guadalajara, Oaxaca, Guatemala, Bogota, Cali, Quito, Trujillo, Lima, Cuzco, Arequipa, Asunción, Buenos Aires, and Santiago de Chile. Between 1540 and 1580, the basis of the Brazilian city system was laid with the founding of Recife, Salvador, São Paulo and Rio de Janeiro in the Portuguese territories. In these years the Spanish city system was further extended — indeed among Spanish American cities only Montevideo and Medellín were founded after the sixteenth century.[1]

The rapid establishment of the Spanish urban system was a response to the desires of the Spanish Crown, Church, bureaucracy and adventurers to control effectively the resources and populations of the Americas. It has been frequently pointed out that a major characteristic of Spanish colonisation has been its geographical attraction to densely-populated areas and the development of complex systems to control these populations.[2] As a result the great majority of Spanish colonial cities were either based on Indian urban centres (Oaxaca, Mexico, Cajamarca, Cuzco, Quito) or founded in regions of dense indigenous populations (Puebla, Guadalajara, Lima, Arequipa). The other important type of urban centre that developed was that which acted as entrepôt between these inland towns and the demands of the imperial power (Veracruz, Panama, Callao, Cartagena).

Thus, the origins of the post-Columbian Spanish American town are markedly different from those of its European and North American counterparts. The European town generally developed as a service centre for an already settled hinterland.[3] The gradual expansion of the North American urban system was closely linked to the westward movement of the frontier.[4] But the town arrived to service rural needs. In Spanish Latin America, on the other hand, 'the town was the source of energy for the occupance, exploitation and administration of the New World'.[5] In other words, the town in colonial Latin America was not a response to the need for urban services by agricultural settlers but was rather the agent of organisation of the Spanish Crown. The colonial town thus contained the religious, administrative, and military functions through which the Spanish Crown effectively controlled and organised the surrounding rural areas. Such a strategy of colonisation did effectively concentrate scarce human resources. By 1580, Spanish America recorded only 23,000 *vecinos* (recognised settlers) but these were already distributed through 189 urban centres.[6]

Town jurisdiction was not restricted to a specific urban area, but included the surrounding countryside up to the boundaries of the next urban jurisdiction. Towns owned their hinterlands both in the sense of economic proprietorship and in the sense of politico-administrative control. These towns became dominated by the *vecinos* who resided in them and who used their municipal authority to consolidate and enlarge their estate. As Morse comments, 'an initial moment of social democracy was therefore followed by consolidation of an oligarchy based on land tenure and prior arrival'.[7]

The sixteenth century was characterised by the vitality of urbanisation in Spanish America. The rapid establishment of a spatially extensive city system allowed the Spanish authorities to create an elaborate urban hierarchy. In Mexico, for example, the hierarchical system of Castile was used (ciudad, villa, pueblo in descending order) for purposes of administration and control. Meanwhile, the Spanish American towns themselves were 'formalised abstractions, constructed on the grid-iron pattern, the rational geometry of empire . . . and obeying the prescriptions of Renaissance town-planners'.[8] Indeed towns had to be built according to the strict specification laid down in Philip II's 'Royal Ordinances for the Laying Out of Towns' of 1573.

However, after the sixteenth century, the urban system in Spanish America virtually stagnated for more than two centuries. This was closely linked to the lack of economic rationale behind such an elaborate city system. Population was too sparsely distributed, and the friction of distance too strong given both prevailing transport technology and terrain. Thus, the development of articulated regional economies fostering urban growth did not occur unless linked to the large-scale mining of high-value metals — as in the Potosí region of Bolivia in the sixteenth and seventeenth centuries[9] or the Mexican Bajío in the eighteenth century.[10] The commercial functions of towns were thus limited outside specific regions, the capitals of Vice-royalties and the ports which linked America to Spain.

The development of urban areas in Brazil contrasts with that of Spanish America due to the less rigid system of Portuguese state control and the development of the plantation as the major nucleus of settlement. As a result, towns developed in response to commercial needs rather than to the dictates of imperial necessity. Thus, the discovery of gold in Minas Gerais in the eighteenth century brought a new wave of urban developments in the central south and the establishment of Rio de Janeiro as the administrative centre of Brazil due to its function as entrepôt for the thriving gold region.

By the end of the eighteenth century, the virtual stagnation of the Spanish American urban system for two centuries had served to preserve intact the urban hierarchies of the sixteenth century and, indeed, had managed to strengthen the local allegiance of urban centres to the major town of the Viceroyalty, audiencia or captaincy-general. It is interesting to note that while the fusion of Hispanic South America under the Spanish Crown was largely achieved through the creation of an elaborate urban system, the disintegration of Hispanic South America into sixteen separate states in the first half of the nineteenth century was linked to the previous dissolution of the unified urban system and its evolution into what has been described as a city-state system.[11] One important factor behind such fragmentation was the friction of distance. In the age before railways, the use of the horse and cart meant that distance was a great brake on all kinds of spatial interaction. For example, in the late eighteenth century, the time it took to transport wine from San Juan to Salta by horse and cart was 40 days[12] (nowadays a lorry would take less than 16 hours). Furthermore, commercial interaction between regions was minimal due to the restrictive measures placed on the regional specialisation of industry by the Spanish Crown. It was also evident that the economic interests of the elites of the major colonial cities such as

Mexico City and Lima were opposed to those of the less important towns, since they had more direct access to crown monopolies and to the other perks of centralisation.[13]

The period in which Hispanic America achieved independence from Spain thus witnessed the disintegration of the continent around the cities of contemporary importance. Between 1810 and 1828, the Viceroyalty of the River Plate divided into four — Argentina (Buenos Aires), Paraguay, (Asunción), Bolivia (La Paz) and Uruguay (Montevideo) and that of New Castile into Peru (Lima) and Chile (Santiago). The Viceroyalty of New Granada lasted somewhat longer due to the political energies and leadership of Simon Bolivar but by 1830 it had nevertheless divided into Ecuador (Quito), Colombia (Bogóta) and Venezuela (Caracas). According to Robinson, with communications still affected by an almost total standstill in certain seasons, the movement of people, products and opinions could not take place from one area to another quickly enough to make so large a unit practicable.[14] Of still more significance was the deep-seated rivalry between Caracas and Bogotá. Caracas, the cradle of liberty, thought itself far more important than the 'Athens of South America'. There was a distinct reluctance in Caracas, as the centre of the revolution, to relinquish ultimate authority to a remote federal centre. Perhaps the most dramatic fragmentation of all took place in Central America. In 1823, New Spain was divided into Mexico and the United Provinces of Central America. Such a union, however, proved untenable due to the rivalries and strife that broke out between the constituent city states. Sixteen years later, Central America divided itself into some of the smallest countries in the world based around the cities of Guatemala, San Salvador, Tegucigalpa, Managua and San José.

Portuguese America, however, remained intact. Singer argued that this was closely linked to the weak administration of the Portuguese authorities and the considerable autonomy that regional elites enjoyed.[15] This was particularly due to the commercial importance of Brazilian cities and their regional, national and international connections. The two northeastern towns of Salvador and Recife constituted the entrepôts for the Northeast's sugar economy while Rio was the commercial hinge for the mineral wealth of Minas Gerais. Even by 1850 these were the three major towns of Brazil, with populations of over 70,000; no other Brazilian town had yet achieved a population of 50,000.[16] Furthermore as these three cities were located on the coast, communication between them was relatively easy.

Urbanisation and the Export Economy

The lack of integration between the major cities of Latin America continued throughout the nineteenth century. The newly-defined Spanish American countries began to reorientate their trading links in favour of northern Europe and the United States. However, each country developed these links individually and trade *between* countries remained stagnant and may even have declined; for example, the colonial wheat trade between Chile and Peru was effectively stopped by high Peruvian tariffs in the post-independence period.

This reorientation of trade came at a time when the industrial revolution was reducing the unit cost of producing manufactured goods and the transport revolution (steamships and railways) was reducing the unit cost of carrying the goods. Trading between Latin America on the one hand, and Europe and the United States on the other, became characterised by a northward flow of raw materials and a southward flow of manufactured products.

In terms of city systems in Latin America, this had at least two effects: the creation of regional urban systems linked to export activities and an increasingly primate city system. From the 1850s onwards, regional urban systems linked to export activities became increasingly articulated by one of the components of the transport revolution, railways. Railways began to be constructed in the 1850s in Latin America and their construction continued apace until the beginning of the twentieth century. Their construction energised both mining and agricultural areas mainly due to the cheaper transport rates, rates that could be as little as a fifth of pre-railway transport costs.[17] The establishment of railway systems created new or fortified existing urban systems. In agricultural areas, such as the coffee lands of Antioquia or São Paulo state, or the wheat lands of the northern Pampas of Argentina, the urban system developed more slowly but has remained more permanent than in mining areas.[18] In mining areas, such as Chile's Norte Chico, railway construction brought rapid urban developments, particularly at the ports where the railway and steamship met. For example, after South America's first railway was built between Copiapó and the port of Caldera in 1852, the latter became 'a boom town, with smelters, water distillation plants, coal storage facilities, docks, warehouses, and housing for more than 2,000 residents within three years of the railroad's construction'.[19] As Figure 8.1 demonstrates

Figure 8.1: Chilean Railways, 1900

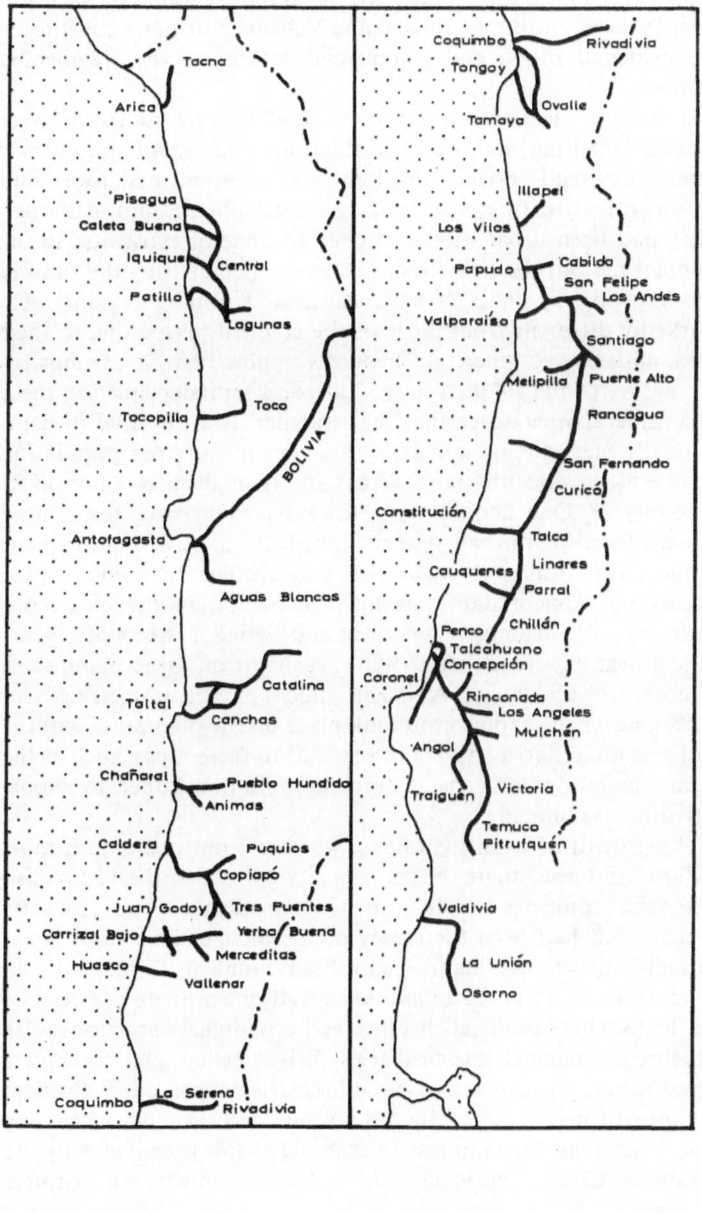

the railway brought a disintegrated, compartmentalised development to the Norte Chico. However, with the exception of the traditional urban centres of Copiapó and Vallenar, the ports constituted the principal urban and commercial centres of the fragmented network.

It was at the ports where the manufacturing functions were located (smelting and water distillation) and where the various commercial and service functions were concentrated (port and railway infrastructure, warehousing and administration). Although there had been urban systems linked to mineral extraction in the colonial period, the new urban systems articulated by the railway and growing up in both agricultural and mining regions were markedly different from those of the colonial period due to their predominantly commercial flavour as opposed to the administrative nature of the colonial town. Indeed, after independence, there is considerable evidence that the 'colonial' town located in commercially stagnant or non-exporting regions suffered population decline given that 'the bureaucracy of colonialism was no longer necessary'.[20] One should not, however, exaggerate the spatial extent of regional urban systems linked to export activities. First, many Latin American countries had neither booming export sectors nor concomitant railway construction in the nineteenth century — for example, Venezuela and Ecuador. Secondly, some agricultural export activities were organised on large plantations adjacent to the coast. As such, much of the processing and packaging of the export crop took place on the plantation and the port was limited to a 'stevedore's role'.[21] In these areas, such as the sugar plantations of coastal Peru, urbanisation linked to export activities was limited.

The growth of export earnings, whether from plantation, mine or farm, did accentuate the primate city pattern of the 15 Spanish American countries created after independence. The city-state systems that had been the legacy of the independence movement normally developed highly centralised political frameworks in order to ensure future stability. As a result, the primate city became the 'locus where political clientage and economic concessions were dispensed, financial intermediaries thrived, elites and foreigners tasted urban pleasures, and basic industries appeared'.[22] Primate city growth was financed by the revenue derived from increasing export earnings. One impressive example of this is provided by the nitrate of Chile's Norte Grande in the late nineteenth century.

Table 8.2: Primacy in Chile, 1813–1920 (population in thousands)

Year	Chile Population	Santiago Population	Valparaíso Population	Index of Primacy	Population of Concepción (3rd city)
1813	583	35	5	7	—
1835	1,033	70	30	2.3	7
1865	1,918	115	70	1.6	14
1875	2,076	130	98	1.3	18
1885	2,527	189	105	1.8	25
1895	2,712	256	122	2.1	40
1920	3,754	507	182	2.8	64

Source: R. M. Morse, Trends and patterns of Latin American urbanization, 1750–1920, *Comparative Studies in Society and History*, 1974, 16, 4 (Sept.), 416–47.

Apart from creating regional urban systems based on the ports of Antofagasta and Iquique, nitrate mining rapidly came to be the principal source of revenue for the Chilean government. In 1895 the revenue derived from the export of over one million tons of nitrate accounted for 66 per cent of Chile's expenditure for that year.[23] Much of this expenditure went to finance the construction of public buildings in Santiago and to extend a railway system based there. As a result, Santiago's primacy, challenged between 1830 and 1880 by Valparaíso's rapid port and commercial growth, was once again reasserted. In 1875 Santiago's population was only slightly larger than that of Valparaíso due to the trebling of the latter's population between 1835 and 1875. After the glorious forty years of Chile's nitrate era (1880–1920), Santiago's population was nearly three times as large as that of Valparaíso (see Table 8.2).

In Peru, the guano prosperity of the 1850s and 1860s dramatically increased the primacy of Lima and its adjacent port, Callao. During colonial times, the combined population of Lima and Callao was less than double that of Peru's second city, Cuzco. However, by 1876, after the guano boom, the population of Lima/Callao was as much as seven times greater than that of Cuzco.[24] Lima's extreme primacy within Peru has been the most notable feature of that country's urban development ever since.

Meanwhile the strength of the mercantile and political elites in Mexico City and the centralisation of the country's railway system maintained Mexico City's primacy throughout the nineteenth century.[25] Furthermore, in Argentina, Buenos Aires recorded consistently high primacy indices in the latter half of the nineteenth century as the Pampas came to provide beef and wheat for the European economies with Buenos Aires the entrepôt for the trade. In the smaller countries of Central America, Bolivia, Paraguay and Uruguay, even higher indices were recorded; in Uruguay, for example, 30 per cent of the country's population lived in Montevideo by 1908.[26]

However, there were three Spanish American countries with no primate distribution — Ecuador, Colombia and Venezuela. McGreevey argues that this was because of the low level of exports that these three countries were able to generate in the nineteenth century.[27] The inability of these three countries to expand significantly their international trade during that century left them with an urban system in which no city had grown substantially more than the rest. Venezuela's discovery and exploitation of oil in the early twentieth century rapidly increased its international trade and its urban structure soon changed. Between 1920 and 1950, the population of Caracas increased sixfold and another primate city had been created.

The evolution of economies increasingly oriented towards export production thus became closely associated with a further centralisation of economies and political activity within the newly-formed countries of Hispanic America. Indeed, there is a close correlation between the success of the export economy and the level of centralisation — although other factors such as the size of country and regional integration also affected the level of primacy. But the combination of export growth controlled by foreign interests and new financially powerful domestic groups, and the evolution of a strong centralised government caused the revenues of export activity to be channelled into the capital city. This was partly because the interests and identities of the emerging economic and political elites overlapped. Roberts argues that 'the development of a strong, centralised state is partly to be explained by the extent to which the dominant classes needed to control and develop central government to expand their economic interests'.[28] Thus as industrial production began in the larger Latin American countries in the early twentieth century and grew more rapidly after the onset of the depression, the process of industrialisation was being engineered upon national territories where the concentration of economic activity, power and population was already well entrenched.

References

1. J. Hardoy (ed.), 1975, *Urbanization in Latin America: approaches and issues.* (Anchor, Garden City, N.Y.)

2. R. M. Morse, 1965, Recent research on Latin American urbanization: a selective summary with commentary, *Latin American Research Review*, 1, 1, pp. 35–74

3. H. Pirenne, 1925, *Medieval cities.* (Princeton)

4. F. J. Turner, 1953, *The frontier in American history.* (New York)

5. A. Portes and J. Walton, 1976, *Urban Latin America: the political conditions from above and below.* (University of Texas Press, Austin)

6. Ibid.

7. R. M. Morse, *Recent research on Latin American urbanization*

8. Ibid.

9. C. S. Assadourian, 1973, Sobre un elemento de la economia colonial: producción y circulación de mercancias al interior de un conjunto regional, *Revista Latinoamericana de Estudios Urbano Regionales, 3,* pp. 135–81

10. D. A. Brading, 1971, *Miners and merchants in Bourbon Mexico, 1763–1810.* (Cambridge University Press, Cambridge)

11. B. Roberts, 1978, *Cities of Peasants.* (Edward Arnold, London)

12. D. T. Halperin, 1970, *Historia contemprānea de América Latina* (Alianza Editorial, Madrid)

13. B. Roberts, *Cities of Peasants*

14. D. J. Robinson, 1971, Venezuela and Colombia, in H. Blakemore and C. T. Smith (eds.), *Latin America: Geographical Perspectives.* (Methuen, London), p. 216.

15. P. Singer, 1973, *Economía política da urbanizaçao.* (Ediçoes CEBRAP: São Paulo)

16. J. Hardoy, *Urbanization in Latin America*

17. R. Miller, 1976, Railways and economic development in central Peru, 1880–1930, in R. Miller, C. T. Smith and J. Fisher (eds.), *Social and economic change in modern Peru.* (Centre for Latin American Studies, University of Liverpool), pp. 27–52

18. R. M. Morse, 1975, The development of urban systems in the Americas in the nineteenth century, *Journal of Interamerican Studies and World Affairs*, 17, 1, pp. 4–25

19. L. R. Pederson, 1966, *The Mining Industry of the Norte Chico, Chile.* (Northwestern University Studies in Geography, Evanston, I11.)

20. B. Roberts, *Cities of Peasants*

21. R. M. Morse, *The development of urban systems in the Americas*

22. R. M. Morse, 1974, Trends and patterns of Latin American urbanization 1750–1920, *Comparative Studies in Society and History*, 16, 4 (Sept.), pp. 416–47

23. H. Blakemore, 1971, Chile, in H. Blakemore and C. T. Smith (eds.), *Latin America: Geographical Perspectives.* (Methuen, London)

24. R. M. Morse, 1974, *Trends and patterns of Latin American urbanization*

25. B. Roberts, *Cities of Peasants*

26. R. M. Morse, *The development of urban systems in the Americas*

27. W. P. McGreevey, A statistical analysis of primacy and lognormality in the size distribution of Latin American cities, 1750–1960, in R. Morse (ed.), 1971, *The urban development of Latin America, 1750–1920.* (Center for Latin American Studies, Stanford University)

28. B. Roberts, *Cities of Peasants*

9 INDUSTRIALISATION AND URBANISATION

It is the purpose of this chapter to explore the modern relationship between industrialisation and urbanisation in Latin America given the urban inheritance that we examined in the previous chapter. However, before pursuing the analysis of this relationship it is worth summarising six of the distinctive features of the Latin American industrial process.

First, industrialisation has traditionally (and certainly since 1929) been promoted in order to supply the national market. Most industrialised and industrialising countries have aimed at supplying the domestic market initially but a subsequent emphasis on exporting manufactured products tends to follow. The propulsive effect of manufactured exports contributed to the industrialisation of the United Kingdom and the European continent in the nineteenth century.[1] The industrialisation of Japan and the newly industrialising countries of East Asia (Korea, Taiwan, Hong Kong, Singapore) have relied even more on export-led growth since the Second World War. However, Latin American industrial expansion, with the possible exception of Brazil's recent drive to increase manufactured exports, has been decidedly inward-looking.

Nevertheless, despite being inward-looking, industrialisation has become the dominant economic force in most Latin American countries displacing in importance agricultural and other primary-producing sectors. Table 9.1 demonstrates the ratio of manufacturing to agricultural Gross Domestic Product (GDP) in 1979 in fifteen Latin American countries. Seven countries have a Gross Domestic Product in manufacturing equivalent to between two and three times that in agriculture. Only five of the smaller countries of Latin America along with Colombia still have agricultural sectors with a GDP greater than that of manufacturing.

Third, the inward-looking nature of Latin American industrialisation during the mid-twentieth century has been associated with an emphasis on consumer-good production. Not only has the emphasis of mid-twentieth century industrialisation worldwide shifted towards more consumer-good production but also, the policy of import substitution has promoted the production of a

Table 9.1: Ratio of Manufacturing to Agricultural Gross Domestic Product, 1979

Country	Ratio
Chile	3.00
Mexico	2.90
Argentina	2.85
Venezuela	2.66
Peru	2.60
Brazil	2.55
Uruguay	2.39
Ecuador	1.27
Costa Rica	1.00
Nicaragua	0.83
Bolivia	0.77
Colombia	0.72
El Salvador	0.54
Honduras	0.53
Paraguay	0.52

Source: World Development Report 1981

full range of consumer goods rather than a more specialised selection.

Such an orientation of production has caused a fourth distinctive characteristic of Latin American industry — a dependence on foreign technology. The production of consumer goods in Latin America had the effect of reducing imports of such goods but the corollary was the need to import the plant and inputs to produce these goods. As plant technique, machine capacity, inputs and product design changed overseas, there was continuous pressure on the Latin American industrialist to change accordingly — in order to maintain his share of the domestic market or to prevent imports if foreign product design had sufficiently changed to warrant renewed imports (as in the case of digital watches). The national firm has tended to be very dependent on foreign technology, machinery and, at least initially, material inputs. The subsidiary of the multinational firm was even more closely linked to sources of foreign technology and inputs.

The reliance on foreign technology and plant has had its impact in a fifth characteristic of Latin American industry — a relatively small generation of employment given the importance of manufacturing to the economy, the size of the labour force and the large numbers of unemployed/underemployed. Indeed, the manufacturing workforce in the modern sector, earning high wages relative to their non-manufacturing counterparts, has often come to form a sort of labour aristocracy — relatively prosperous but small in number. It is generally accepted that the import of capital intensive plant and machinery has been distinctly out of step with the labour market in Latin America.

A sixth characteristic of Latin American industry is its spatial concentration. National manufacturing production, employment and value-added are commonly concentrated in one, sprawling, multi-centred metropolitan area. Despite the application of many government strategies designed to reduce such concentration, decentralisation of manufacturing activity has only been successful at relatively short distances from the primate city — a process that can best be seen as a further strengthening and intensifying of the process of centralisation.

What implications has such a distinctive form of industrialisation had on the process of urbanisation? Initially, it would be appropriate to examine two general points. First, there has not been such an intimate connection between industrialisation and urbanisation as occurred in nineteenth-century Europe. Urban growth in the UK and continental Europe closely followed industrial growth and the expansion of employment. In Latin America, however, increases in manufacturing employment have not been nearly so dramatic as the growth in urban dwellers. The growth in urban dwellers, caused by high urban fertility rates and large flows of migrants from rural areas, has far outweighed any increases in manufacturing employment. In Europe, surplus urban populations had the option of international migration, an opportunity denied to the equivalent populations in the modern Latin American city. Secondly, industrialisation has reinforced the spatially concentrated pattern of urbanisation in Latin America. The last chapter demonstrated that primacy became a distinctive feature of the urban geography of most countries during the export economy phase of Latin America's growth during the nineteenth century. For example, by the turn of the century both Peru and Argentina had primate cities with populations more than seven times as large as their second cities. However, by 1980, after eighty years of industrialisation, Lima and Buenos Aires were more than eleven times larger than their respective second cities.

In the early phase of industrialisation, entrepreneurs and firms had been attracted by the large concentration of population provided by the primate city and the location of economic and political power within those cities. But the relationship between primacy and industrialisation became mutually reinforcing. Industrialisation in turn increased employment in the primate city, both directly, in terms of the expansion of the manufacturing work force, and indirectly, in terms of the growth of related employment in the public and private services. Furthermore, the process of industrialisation created diverse but powerful elites in the primate city. Families who had made fortunes from industrial enterprises, often from recent immigrant stock, came to live there. Spatially concentrated industrialisation also provided the possibility for trade union organisations to flourish — even in apparently repressive states. Economic activities related to industry, such as banking and commerce, also became concentrated in the primate cities alongside their respective elites.

Government, itself normally located in the primate city, became more open to the influence of a wide variety of industrial pressure groups. Each industrial sector created its own representative association and pressure group — such as ASIMET, that represents and pressures for all metal-working industries in Chile. Companies from overseas lobbied collectively through chambers of commerce named after their country of origin. Large national firms organised themselves through well-funded, prestigious organisations such as SOFOFA in Chile. Small- and medium-sized firms attempted to lobby and improve conditions through wider sectoral groupings. Although these organisations were diverse in nationality, sector and size of firm, they had one thing in common — their representative bodies were located in the primate city. As a result, governments became more responsive to the requirements and needs of industrialists and particularly those located in the primate cities. Cheap urban food policies that favoured both the industrial entrepreneur and worker (but not the agricultural sector) were a result of such government partiality to industry and the primate city in the 1950 and 1960s.

Industrialisation and the Urban Hierarchy

The spatial association of industrialisation with the primate city has had the corollary that towns further down the national urban hierarchy have attracted little industry. In particular, the inward-

Table 9.2: Maufacturing Workers as a Percentage of the Total Non-agricultural Workforce for Urban Areas in Chile, 1970

	Total Non-agricultural Workforce	Manufacturing Workforce	B as a % of A
Santiago	848,606	199,729	23.5
Concepción	113,014	28,727	25.4
Antofagasta (North)	33,516	3,499	10.4
Temuco (South)	30,431	4,442	14.6
Rancagua (Central Valley)	24,682	2,389	9.7
Chillán (Central Valley)	23,691	3,035	12.8
Puerto Montt (South)	19,074	2,444	12.8
Los Angeles (South)	14,765	1,819	12.3
Coquimbo (North)	12,564	1,251	10.0
Curicó (Central Valley)	12,172	1,353	11.1
Small towns in Santiago Province			
San Bernardo	31,504	7,069	22.4
Puente Alto	21,600	6,622	30.7
Peñaflor	8,638	3,050	35.3

Source: Instituto Nacional de Estadisticas, *Caracteristicas Basicas de la Población, Censo 1970*, Santiago, 1973

looking, consumer-good industries were strongly attracted to the large city and avoided the smaller provincial towns. The need to be near the major internal market and to have easy contacts with foreign technology and companies were always vital considerations for these industries. As a result, manufacturing has had relatively little impact on the occupational structures of small- and medium-sized towns in Latin America. Table 9.2 demonstrates this relationship in the Chilean urban hierarchy. Santiago and the port-industrial town of Concepción had approximately one-quarter of their substantial workforces in the manufacturing sector. Meanwhile, medium- and small-sized towns away from Santiago had only about one in eight of their workers employed in manufacturing. However, the attraction of industry for a location near the primate city can be gauged from the differential performance of medium- and small-sized towns located near Santiago. San Bernardo, Puente Alto and Peñaflor had up to 35 per cent of their workforce involved in manufacturing (see Figure 9.1).

Thus, the precise relationship between industrialisation and urbanisation depends not only on the size of town but also on its proximity to the major city. Evidence from various parts of Latin

Figure 9.1: Chile's Central Region

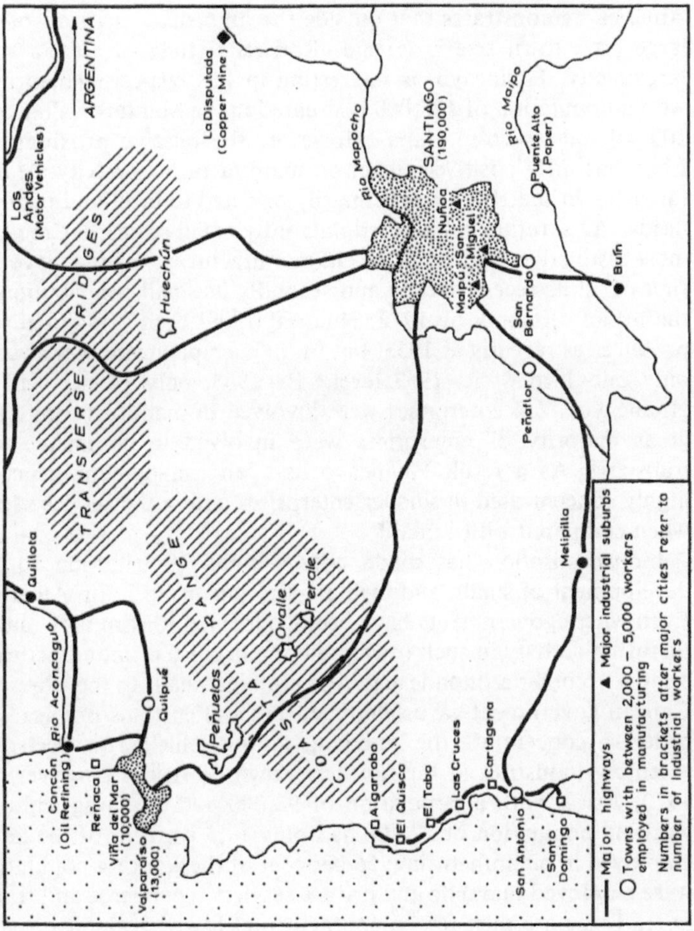

America demonstrates that outside the immediate environs of the large city, town size is crucial. Roberts's study of Peru's sixth largest city, Huancayo, is interesting in this respect.[2] Huancayo, with a population of 125,000 is situated in the Mantaro valley only 200 kilometres from Lima. However, the relative proximity of Lima has little positive impact on manufacturing activity. Manufacturing in the 1970s was primarily oriented to local and regional needs. As a result it was predominantly small-scale. But the two most distinctive features of recent manufacturing in Huancayo are first of all its recent decline and secondly its small contribution to the urban economy by 1972. Huancayo had been a major textile producer as recently as 1953, but by 1965 employment was already only one-third of its 1953 level.[3] By 1965, only 14 per cent of Huancayo's 214 enterprises were involved in manufacturing. The great majority of enterprises were involved in commerce and transport. As a result Huancayo had 'an employment structure highly concentrated in smaller enterprises and in the service sector when compared with Lima'.[4]

Industrialisation has made only a small contribution to the development of small- and medium-sized towns in Latin America. Even where governments have been concerned to stimulate manufacturing activity in such towns, the importance of manufacturing relative to other economic activities remains small. In the 1960s the Chilean government gave significant tax and customs privileges in order to concentrate the emerging motor vehicle and electronic assembly industries in the northern town of Arica. However, by 1970, Arica, with a population of 92,000 and a non-agricultural working population of 27,700 had only 19.5 per cent of the latter employed in manufacturing; 56 per cent of the working population were employed in public and private services, commerce and transport. There are then few 'manufacturing' towns in the lower and middle ranges of the urban hierarchy — unless located near the primate city. The proportion of the work force involved in manufacturing is consistently lower than that in the large cities and agglomerations.

Industrialisation in the Large City

What then is the relationship between industrialisation and the growth of the large city and sprawling agglomerations? First of all, the relationship has been and is much more intimate than in the smaller city. However, the precise relationship tends to depend on the more general economic and political context within which industry develops.

Cardoso and Faletto distinguish between three political and economic environments within which industry expanded after the export economy phase of Latin American development.[5] One can summarise these as follows:

1. Industrialisation undertaken primarily by a national bourgeoisie and without significant economic intervention by the state — the case of Buenos Aires and Argentina.
2. Industrialisation undertaken by a regional bourgeoisie but which benefited from the active support of the state — the case of São Paulo and Brazil.
3. Industrialisation largely promoted by the state with the national bourgeoisie playing a more passive role — the case of Mexico, Chile and the smaller countries of Latin America.

It is, of course, hazardous to make such sweeping generalisations about the economic history of Latin America and detailed economic histories such as those of Glade make such distinctions open to question.[6] However, it is generally accepted that an emerging industrial bourgeoisie and the state were the main institutional agents of early twentieth century industrialisation in Latin America. Furthermore, the precise nature of the contributions of these two agents of early industrialisation had considerable impact on the contemporary and subsequent patterns of urbanisation.

European immigrants played a prominent role in the rise of an industrial bourgeoisie. The traditional land-owning classes of Latin America generally had little interest in urban-based manufacturing and other groups (e.g. urban professionals) were deterred by the risky nature of manufacturing investment. Immigrant entrepreneurs from Europe were more inclined to face the risks of initiating manufacturing and certain factors were in their favour. First, access to capital from European banks and trading companies was relatively easy for immigrants. In São Paulo, 'the immigrants appeared to the European trading companies to be the most dependable instruments for the advancement of their interest'.[7] Secondly, some immigrants had a practical knowledge of a relevant industrial technology — either from their work experience in

Europe before migration or from working in an immigrant enterprise in Latin America. With favoured access to both capital and technology and with an appreciation of the relative risks and benefits of manufacturing, the European immigrant became a prominent member of the industrial bourgeoisie in many Latin American countries.

As Cardoso and Faletto noted, it was Buenos Aires and São Paulo that attracted the largest number of European immigrants.[8] A combination of booming export economies (beef, wheat and coffee) and well-financed government immigration policies provided both means of passage and work on arrival for the immigrant. Many immigrants preferred to stay in the major city rather than work in the agricultural hinterland. The result was fast-growing cities. São Paulo's population grew from 23,000 in 1872 to 58,000 in 1920; of the latter figure, two thirds were foreign-born or their offspring.[9] The population of Buenos Aires increased from 300,000 in 1880 to 1.5 million in 1914, of whom 50 per cent were European born.[10]

It was within these large concentrations of European immigrants that the immigrant entrepreneurs initiated and expanded their manufacturing operations. Dean notes that few European immigrants settled in the Brazilian Northeast and that immigrant entrepreneurs did not develop large and successful businesses here.[11] In São Paulo, however, the potential entrepreneur had access to European trading capital, European technology, skilled immigrant labour and a reasonably large market for European manufactured goods. In fact, many industrialists started off as importers of manufactured goods before identifying the greater opportunities provided by manufacturing.

The impact on urban growth was profound in both the short and long term. Immigrant entrepreneurs shifting from importing to manufacturing were involved in greater risks although with greater potential for profit. Related to this, their strategy was often one of minimising risk. In locational terms, this meant building up their industrial enterprise in the 'immigrant city'. In organisational terms, it could signify aiming for vertical linkage and assuring the supply of inputs and services. In this respect, an interesting case is provided by the building up of the Matarazzo industrial empire in São Paulo in the early twentieth century.

After a brief career in importing in São Paulo, Francisco Matarazzo built a lard factory in the 1890s (he had had a similar business in Italy before emigrating). The new product, produced and sold more cheaply than imports, swept foreign competition from the market. In the early 1900s he diversified into flour milling and built three mills in São Paulo. He decided to make the sacking for the flour, and erected a cotton mill with printing machines in 1904. Rather than buy cotton from the ginners, he installed machines of his own and bought the cotton in the boll.[12] The valuable cottonseed was pressed and used to make soap and glycerine. Since the lard business and the flour milling depended on raw materials from Argentina, Matarazzo bought some coastal freighters. Later he bought another cotton mill, a cannery, a lithography shop, a box plant (linked to a sawmill), foundries and machine shops to repair his equipment. He built his own docks for the coastal fleet (due to the pilferage and slowness of other docks), kept a special contingency fund so that he would not have to buy insurance and banked through the branch of the Bank of Naples that he directed.

Matarazzo and other immigrant industrialists were interested in vertical integration in order to reduce risks and the uncertainties of raw material supply, transport and power in a frontier economy. In so doing they effectively created a series of industrial 'growth poles' reminiscent of Perroux's theory (see Chapter Seven). The important implication for urban growth was that these 'growth poles' were all concentrated in the city in which the enterprise had originated — in Matarazzo's case, São Paulo. Indeed, Matarazzo bought a large section of São Paulo's Agua Branca district and grouped most of his industrial plant there.[13] Immigrant entrepreneurs thus developed industrial enterprises in cities already growing through the settling of large numbers of European immigrants. In turn, the increased industrial activity stimulated a greater need for labour and further immigration. A process of circular and cumulative growth had begun.[14] In the longer term, the immigrant entrepreneurs left an industrial infrastructure and complex system of industrial linkage that would attract more and more industry in subsequent periods.

Two further points should be made here. First, although immigrant entrepreneurs were dominant in the emerging industrial bourgeoisies of São Paulo and Buenos Aires, other groups (such as landowners) also participated in manufacturing. Secondly, immigrant entrepreneurs were important in promoting industrial enterprises in other cities of Latin America. In Santiago, one study found that 76.5 per cent of all industrial enterprises were set up by immigrants.[15] The 1928 Chilean industrial census showed that 23 per cent of industrial entrepreneurs were of immigrant origin, and of these 60 per cent had decided to establish their plant in either Santiago or Valparaíso.[16] On the other hand, the presence of few immigrant entrepreneurs has been seen as one factor explaining the lack of early manufacturing activity in such Latin American cities as Lima (where industrial growth was small between 1900 and 1940).[17] However, in order to analyse the relationship between industrialisation and the large city in other countries in Latin America, reference must also be made to the impact of the state.

According to Cardoso and Faletto, the state was the main force behind industrialisation in Latin American countries other than Argentina and Brazil.[18] There is a chronological point to be made here. The emergence of an industrial bourgeoisie closely linked to immigrant entrepreneurs stimulated the industrial and urban growth of Buenos Aires and São Paulo between 1880 and 1930. The state however became more prominent in stimulating industrial and urban growth after the world depression that began in 1929 (see Chapter Two). Cardoso and Faletto argue that the state became prominent in the countries whose export economy phase had not been in the hands of national entrepreneurs (beef 'estancieros', wheat tenant farmers, coffee planters) but constituted an enclave, controlled by foreign interests and capital, such as the nitrate and copper mines of Chile.[19] In these cases there was not the potential for an easy shift of capital from primary to secondary production, and manufacturing was additionally hampered by the small internal market that the enclave economy did little to expand.

Cardoso and Faletto point to Chile as an example of the state sponsoring the beginnings of industrialisation when faced with the crisis in the enclave economy. In fact, as Palma has shown, considerable industrial development had occurred before 1930 in Chile, and immigrant entrepreneurs had been prominent in these developments.[20] However, between 1928 and 1937, with the Chilean state giving greater protection to domestic industry, employment in Chilean manufacturing increased by 50 per cent. Specific policies designed to transfer capital from the export enclave to manufacturing date from 1939 when a government holding company, the Chilean Development Corporation (CORFO) was created, financed by a tax increase on copper mining. Within the next two decades, CORFO granted loans to hundreds of enterprises in a wide variety of industries and made investments in the stock of some 66 companies, in 41 of which it held a majority interest.[21] Considerable effort was devoted to promoting investments in peripheral regions in such schemes as hydro-electric projects, oil exploration, production and refining, beet sugar refineries and the Huachipato iron and steel mill. But many investments were also made in Santiago and some, such as MADECO's copper mill and wire plant, would have been well suited to a decentralised resource location. Furthermore, the whole administative machinery which governed Chilean industrial growth — price supports, subsidies, new sectoral initiatives — evolved within the expanded state bureaucracy of Santiago.

State intervention in the industrial process occurred in most Latin American countries in the decades of the mid-twentieth century. Peron's period of power (1946–55) in Argentina saw government promoting a rapid process of industrialisation based on national entrepreneurship. Positive action by the state to further domestic industrialisation also became the keynote of the *Estado Novo* programme espoused by the long regime of Getulio Vargas who ruled Brazil from 1930 to 1945 and again from 1950 to 1954. The experience of Venezuela since the first *Acción Democratica* government acceded to power in 1945 has been similar with the creation of a government bank for industry and a national development authority absorbing petroleum earnings and reinvesting them in industry.

Industrialisation created employment for the rapidly-expanding large cities of Latin America. In the 1930s and 1940s, large city growth in Argentina, Brazil and Chile was accomplished by substantial increases in employment in the manufacturing sector. In Santiago, manufacturing employment more than trebled between 1928 and 1957 — from 38,000 to 124,000. It has been argued that as a result most of the urban classes of the large cities derived advantages from the process of industrialisation. 'The middle classes received profits and high salaries from the increase in industrial productivity and the increasing demand for commercial and other services which it entailed. The working class, many of whom were of rural or of immigrant origin, could also feel themselves to be in a better situation.'[22]

The relationship between industrialisation and urbanisation in Latin America has therefore been closest in the large city and particularly in those large cities in which European immigration has been prominent. State involvement in the process of industrialisation has further favoured the growth of the large city particularly the capital cities with their concomitant bureaucracies. However, many capital cities, such as Santiago, have only between 25 and 33 per cent of their work force in manufacturing. Many cities attracted large numbers of migrants from the provinces, too many for the opportunities provided by manufacturing industry. With the decline of import substitution as an industrial policy (see Chapter Two) some large cities such as Santiago have recently recorded declining employment in manufacturing. Even in the large city, then, the relationship between industrialisation and urban growth is not universally close. Have any models been developed which attempt to conceptualise this idiosyncratic relationship between industrialisation and the growth of the large city in Latin America? The models most commonly referred to are those which conceptualise the economies of the large city of Latin America as consisting of a dual system.

The Urban Dual Economy

The model most commonly used to examine the impact that capital-intensive industrialisation has had on large cities with rapidly-expanding populations is that of the urban dual economy. This model portrays the urban economy as divided — basically between a capital-intensive sector and a labour-intensive sector. A wide variety of terms have been used to categorise the division — the firm versus the bazaar economy,[23] the upper against the lower circuit,[24] the large-scale versus the small-scale sector,[25] the formal versus the informal sector.[26] All categorisations see the urban economy as dichotomous. A dividing line between the two economies can be conceptualised. It is further argued that such dichotomies are self-sustaining and will persist into the foreseeable future. The small-scale sector, according to Roberts, produces 'those goods for which the market is so reduced and so risky that large-scale enterprises are not interested to enter. Consequently, the small-scale sector is left with activities of low profitability that do not permit capital accumulation, making it unlikely that enterprises in this sector will develop into large-scale ones'.[27]

The concept is an appealing one. In any large Latin American city, there is the contrast between the modern, capital-intensive factory and the sweatshop, the supermarket and the street vendor. But the crucial question must be whether the complex urban economy of the large Latin American city can be conceptualised around such extremes? To borrow Brookfield's phrase, does the use of a dichotomy in this instance 'bedevil clarity of thought'?[28]

Santos delves deepest into the two circuits of the urban economy. Indeed, his book, *The Shared Space*, is completely devoted to the concept. His definition is representative of many definitions of the urban dual economy:

> the upper circuit consists of banking, export trade and industry, modern urban industry, trade and services, and wholesaling and trucking. The lower circuit is essentially made up of non-capital intensive forms of manufacturing, non-modern services generally provided at the "retail" level and non-modern and small-scale trade.[29]

Most of the book is devoted to exploring the characteristics of the two circuits. Santos tabulated the major characteristics of the two circuits and Table 9.3 reproduces them. The dichotomous nature of the urban economy is evident. There is the sharp and consistent division between the capital-intensive and capital-abundant upper circuit and the labour-intensive and labour-abundant lower circuit. In the upper circuit, there are regular wages, large inventories, fixed prices, high fixed costs, government aid and advertising; in the lower circuit, none of these. In the lower circuit, personal relations are the norm — for credit, with clients, over prices. In the upper circuit, a more impersonal system predominates. That such a typology represents two extremes of the urban economy in Latin America, there is no doubt. That such a typology can be extended to justify a dichotomous urban economy in Latin America seems more open to question.

Any dichotomy is arbitrary but its greatest flaw is that it shifts analysis away from the interface. The urban dual economy model has generally been used as a framework within which to analyse the lower circuit or small-scale economy. Analyses have generally assumed that firms within the small-scale economy *stay* there.[30] Capital accumulation, it is assumed, is too small for firms to make the transition from small to large-scale. Perhaps as a result, the focus of research in the small-scale or informal sector has been on enterprises that are firmly rooted in the small-scale/informal sector

Table 9.3: Characteristics of the Two Circuits of the Urban Economy in Latin America

	Upper Circuit	Lower Circuit
Technology	capital-intensive	labour-intensive
Organisation	bureaucratic	primitive
Capital	abundant	limited
Labour	limited	abundant
Regular wages	prevalent	exceptionally
Inventories	large quantities and/or high quality	small quantities poor quality
Prices	generally fixed	negotiable between buyer and seller (haggling)
Credit	from banks, institutional	personal, non-institutional
Profit margin	small per unit; but with large turnover considerable in aggregate (exception = luxuries)	large per unit; but small turnover
Relations with customers	impersonal and/or on paper	direct, personalised
Fixed costs	substantial	negligible
Advertisement	necessary	none
Re-use of goods	none (waste)	frequent
Overhead capital	essential	not essential
Government aid	extensive	none or almost none
Direct dependence on foreign countries	great; externally orientated	small or none

Source: M. Santos, *The shared space: the two circuits of the urban economy in underdeveloped countries* (Methuen, London, 1979), p. 22

— such as street traders, or garbage pickers. Little research has been carried out on the small-scale manufacturing sector in large Latin American cities. Santos, for example, makes no analysis of or reference to manufacturing in the lower circuit of the urban economy.

When reference is actually made to small-scale manufacturing, the value of the urban dual economy model can be brought into question. This point can be best made referring to a study of the Caracas *biblioteca* industry in 1979.[31] According to Morris, 30 per cent of the population of Caracas live in flats in high-rise residences.[32] The majority of new construction in the 1970s consisted of high-rise residences. Owing to the tightly-packed nature of such residences, considerable demand developed for space-saving furniture, in particular modular furniture — modules of shelving and cupboards attached to walls and linking in with tables, bed-bases, and so on. Such furniture was exclusively made in Venezuela (due to high tariffs) but, by using a production process based on chipboard and formica-laminate pressing, such products could be made cheaply. Two well-established furniture firms diversified into the area producing for existing retail outlets. But the most interesting development (from the point of view of the argument) was the proliferation of small entrepreneurs that entered into the market in the mid-1970s. These enterprises had many features of the so-called lower circuit but at the same time had some significant characteristics of the upper circuit. In order to substantiate these points it is necessary to examine one of these small 'biblioteca' enterprises in greater detail.

'Decoraciones Gaete' had many of the aspects of the small-scale economy. First, it was located in the squatter settlement of Mariperez. In Caracas, illegal land seizures in the 1950s and 1960s resulted not only in subdivision of land for housing (ranchos) but also for enterprises (estacionamientos). As a result, overhead capital was small — furniture was made and sold in a building built (in the 1960s) of breeze-blocks and corrugated-iron roofing. Referring to Santos's tabulation, government aid was non-existent, rent was paid to the owner of the plot and fixed costs were small. Thus far the characteristics are those of the lower circuit.

But the enterprise had other characteristics. In order to sell its product to the flat-dwellers of Caracas, it advertised every Sunday in the colour supplement of a leading newspaper. Clients would either phone for the visit of a salesman or call in to the 'estacionamiento'. As prices were much lower than in the traditional retail outlets, demand steadily increased between 1975 (when the enterprise started) and 1979. In 1978, such was the congestion that another building was rented in the squatter settlement of Petare. Mechanisation was important to the operation with furniture saws, presses (for the chipboard and formica), and power drills. Twenty workers were employed in 1979, each receiving regular wages. Credit from the bank was tight but available, a factor that operated to keep inventories low.

The importance of examining this case study is that it points to two major deficiencies of the urban dual economy model. First, the assumption that manufacturing enterprises will have the characteristics of either the upper or lower circuits is strongly refuted by the case study. Secondly, the assumption of all dichotomies that the interface between the two groups is relatively unimportant seems in this case to be untenable. The 'biblioteca' enterprises were clearly at an interface between the upper and lower circuits of the urban economy. In a steadily expanding market, most were growing rapidly, investing in small machinery, taking on more labour, extending their premises. Indeed, such behaviour can be compared with small manufacturing enterprises in East Asia or Western Europe. In these areas, it is generally accepted that the small manufacturing firm adds considerable dynamism to the manufacturing sector — through producing more cheaply, introducing new products, designing or inventing new products, identifying areas of demand that larger enterprises have overlooked. Watanabe (1970), for example, attributed the rapid industrialisation of Japan to the widespread entrepreneurship which could unfold in the large number of small-scale enterprises.[33]

The dynamic nature of small manufacturing firms in Latin America is substantiated by the few studies that have focused on such a topic. Schmitz's Brazilian case studies of knitting and clothing manufacturing in Petropolis, the hammock industry of Fortaleza and the weaving industry of Americana analyse the entrepreneurial nature of small-scale manufacturing activity.[34] He concludes that workers set up their own enterprises not as a survival strategy but in order to become upwardly mobile. 'Probably the single clearest finding which emerges in all three case studies is that the small-scale producers are not unsuccessful job seekers, but rather the contrary: they tend to be skilled workers who have left their jobs of their own accord' . . . 'setting up their own business gives a better chance for economic and social advance than wage employment'.[35] Schmitz further takes issue with the concept put forward by the International Labour Office that small-scale manufacturing is easy to enter.[36] Entrants into the three industries that he studied were invariably skilled workers with previous knowledge of the manufacturing activity and needed significant amounts of capital — between US $1,100 and US $3,600 — or the equivalent of between eight and fourteen months' wages.[37] Peattie's work on the Colombian shoe industry further suggests that in small-scale manufacturing the hypothesis of upward social mobility is more likely than the survival hypothesis.[38]

Perhaps another reason for the popularity of the urban dual economy model was that it provided a simple framework within which to explore the impact of capital-intensive industrialisation on urban areas whose populations were increasing regardless of job opportunities in manufacturing. In this context, the small-scale or informal sector was supposed to act as the refuge for surplus labour. The preceding analysis has demonstrated that small manufacturing enterprises have not acted as a refuge for labour, and as entrepreneurs have needed skills and capital to start such enterprises have decided to adopt such a strategy due to the potential for upward mobility. In this sense, the labour surplus has only had an indirect bearing on small-scale manufacturing firms as its existence contributed to low wages (in the formal sector) which in turn increased the desire to leave employment and to establish a small enterprise.

The urban dual economy model has looked at the two extremes of the urban economy. These may be significant and large extremes in terms of the number of enterprises inhabiting them, but they nevertheless represent only part of the urban economy. That such a model has developed such wide acceptance can be attributed to its rapid and enthusiastic adoption by the International Labour Office in 1972[39] and to a more general characteristic of human perception that Brookfield describes as follows: 'the world looks simpler in binary, and social science and the development field are no exceptions'.[40] More controversially, one could add that the model served to divert attention away from small manufacturing firms because such firms patently did not 'fit' into the constraints of the model. As more research is being carried out on small manufacturing firms, the failings of the model are becoming more evident even to the still large number of its exponents.

Industrialisation and Urban Marginality

A major criticism of the urban dual economy model was that it analysed two extremes or poles of the urban economy and formulated a dichotomy from these two poles. In so doing, it ignored aspects of the urban economy intermediate to these two poles, and most notably small-scale manufacturing industry. Marginality theory has avoided such problems by concentrating on the 'marginal pole' of the economy.[41] The 'marginal' pole of the urban economy is seen as a response to capital-intensive industrialisation, on the one hand, and a rapidly-expanding urban population, on the other. The secondary sector, Quijano argues, 'has no possibility of absorbing the greater proportion of the urban workforce. The displacement of manpower from rural to urban sectors of the economy does not consist then in an exchange of labour between primary and secondary sectors, except to a minor extent, since the major part of the displaced labour force does not enter the secondary sector'.[42] Furthermore, industrialisation can displace workers, as when new technology reduces the labour needs of an industry. The increased productivity of new industries displaces workers in traditional sectors. Artisan production is increasingly destroyed as machine-made cloth replaces the products of domestic looms and as plastic and aluminium utensils replace earthenware. Traditional food and drink industries give way to modern canning and bottling plants. As a result, these processes create a substantial surplus labouring population.

Quijano's concept of the marginal pole of the Latin American urban economy is bleak and sombre. Capital-intensive industrial development is characterised as being foreign-owned and divorced from national needs apart from those of the upper and middle classes. The growth of national entrepreneurial activity in medium and small-scale manufacturing is seemingly dismissed due to a lack of historical precedent.[43] The pre-eminence of 'monopoly' capitalism, or the domination of foreign multinational corporations, does not permit the emergence of what Quijano terms 'competitive capitalism', or national firms competing for national and foreign markets. Due to this deep-seated structural dominance of 'monopoly' over 'competitive' capitalism, the process is irreversible — monopoly capitalism will continue to extend its power and the marginal pole of the economy will steadily expand.

What is the 'marginal pole' of the Latin American urban economy? Quijano defines it in terms of those economic sectors in which the urban populations marginalised by capital-intensive industrialisation seek work. These sectors are primarily tertiary, such as petty trade, personal services and small repair shops. Quijano makes the important point that such marginal employment is likely to concentrate in the large cities of each underdeveloped country. Earning opportunities are greater in those centres given the concentration of high-income earners. For example, an average upper-middle class house owner in Santiago or Lima may employ a whole range of people in an average year —

two or three domestic servants, a gardener, an odd-job man, a painter — the low wages or economical contracts of such full- or part-time jobs make such employment very common. Apart from employment, the large city can normally offer the 'marginalised' population a greater range of both educational and welfare facilities.

If high unemployment and 'make-work' schemes are statistical indicators of a marginal pole, some large cities in Latin America already have massive marginalised populations. At the end of 1982, Santiago had, according to official figures, an unemployment rate of 25 per cent; in addition, 'make-work' programmes accounted for a further 7 per cent of the 1.3 million workforce. Of the unemployed, 41 per cent registered their previous employment as being in either industry or construction; only 20 per cent recorded their previous employment as being in communal, social and personal services.[44] Such data seem to indicate that during a recession unemployment is felt more acutely in capital-intensive industrial sectors than within the tertiary sectors.

Such an interpretation would support Quijano's contention of a marginal pole becoming steadily larger and sectorally and spatially more confined. As capital-intensive services such as supermarkets, car-washing machines, modern automobile repair services and other labour-saving enterprises develop, more labour will be displaced into the marginal pole. At the same time, there will be fewer links between the marginal pole and the more wealthy urban classes. The marginalised population will turn in on itself and the marginal economy will specialise in producing goods and services for the poor. Such a sectoral development of the marginal pole will have evident spatial implications as squatter settlements will become the spatial confine for such a marginalised economy. In this way, the marginal pole becomes afunctional in terms of the wider urban economy.

Quijano's concept of capital-intensive industrialisation and the gradual emergence of an afunctional marginal pole does have some flaws. First, there is the problem that the marginalised population does well in the large city due to the casual employment available linked to the high earnings enjoyed by wealthier groups; however, it is argued that despite these links being reduced with time, the marginalised populations will stay in the large city. More critical, however, is the question of the reduction of linkages between the 'marginalised' population and the urban economy. Returning to the example of the Santiago house-owner, it is doubtful whether he would wish to reduce his links with cheap domestic and household services — even if he does purchase washing-machines and other labour-saving devices. Indeed if labour became cheaper, he might even wish to increase his use of domestic service.

In a more general sense, Quijano underestimates capitalism as 'an internally expanding force'[45] and its capacity to transform the social and economic structures of urban areas. In particular, Quijano underestimates the potential of employment generation of modern industrialisation. Roberts demonstrates that from 1950 to 1970, the transformative sector (manufacturing and construction) added some 1.63 million new jobs to the Brazilian economy at an average annual increase of 4 per cent.[46] Meanwhile, the tertiary sector (distribution, personal and industrial services, social services) added 6.78 million jobs at a rate of 4.8 per cent a year. Furthermore, from 1960 to 1970 the secondary sector grew at a faster rate than the tertiary sector. 'These figures question the generality of Quijano's marginality hypotheses since those sectors most closely linked to capital-intensive industrialisation appear to be among the most dynamic in creating job opportunities; there is little sign of capital-intensive industrialisation displacing workers'.[47]

According to such Brazilian marginality theorists as Kowarick[47] and Faria,[49] the fact that industrial enterprises are both concentrated in large cities and substantial generators of employment means that the marginal pole or the marginalised economy will be more apparent in provincial rather than in the large cities. For example, in Rio de Janeiro and São Paulo an average of 36 per cent of the workforce operates in manufacturing while in the northern provincial cities of Recife, Salvador, Belem and Fortaleza, an average of only 26 per cent does. These differences in employment are also reflected in differences in mean family income, industrial productivity and *per capita* expenditures on social services — all serving to reduce the size of the marginal pole in the large city *vis-à-vis* the provincial city.

It is worth pointing out that Quijano's concept of the marginal pole was strongly influenced by his Peruvian experience and that critiques of his work by Brazilian theorists naturally reflected that country's background. Referring back to Chapter Two, it was noted that industrialisation had been most successful and far-reaching in the larger countries of Latin America and most especially Brazil. Meanwhile, small- and medium-sized countries, particularly when they followed policies of import substitution, tended to generate a much less dynamic manufacturing sector. Quijano's model tends to fit better into those countries that have achieved only a limited industrial growth, whose industry is dominated by foreign corporations and whose middle class is relatively small. At the same time, the critiques of Quijano's model tend to be most forceful when applied to rapidly industralising countries, such as Brazil, Mexico or Venezuela.

References

1. C. Trebilcock, 1981, *The Industrialization of the Continental Powers, 1780–1914*. (Longman, London)
2. B. Roberts, 1976, The social history of a provincial town: Huancayo, 1890–1972, in R. Miller, C. T. Smith and J. Fisher (eds), *Social and economic change in modern Peru*. (Centre for Latin American Studies, University of Liverpool)
3. Ibid., p. 156
4. B. Roberts, 1978, *Cities of Peasants*. (Edward Arnold, London), p. 86.
5. F. H. Cardoso and E. Faletto, 1979, *Dependency and development in Latin America*. (University of California Press, Berkeley), pp. 74–126
6. W. Glade, 1969, *The Latin American economies: a study of their institutional evolution*. (Van Nostrand, New York), pp. 419–82
7. W. Dean, 1969, *The industrialization of São Paulo*, p. 56
8. F. H. Cardoso and E. Faletto, *Dependency and development in Latin America*
9. W. Dean, *The industrialization of São Paulo*, p. 51
10. J. R. Scobie, 1971, *Argentina: a city and a nation*. (Oxford University Press, New York), pp. 131–3
11. W. Dean, *The industrialization of São Paulo*, p. 53
12. Ibid., p. 62
13. Ibid., p. 63
14. A. Pred, 1977, *City-Systems in Advanced Economies*. (Hutchinson, London), p. 90
15. J. Petras, 1969, *Politics and Social Forces in Chilean Development*. (University of California Press, Berkeley), p. 39.
16. R. N. Gwynne, 1978, *Industrial Decentralisation in Chile*. (Unpublished doctoral dissertation, University of Liverpool), p. 68
17. R. Thorp and G. Bertram, 1976, Industrialisation in an open economy: a case study of Peru, 1890–1940, in R. Miller, C. T. Smith and J. Fisher (eds), *Social and economic change in modern Peru*. (Centre for Latin American Studies, University of Liverpool), p. 62
18. F. H. Cardoso and E. Faletto, *Dependency and development in Latin America*
19. Ibid.
20. J. G. Palma, 1979, *Growth and Structure of Chilean Manufacturing Industry from 1830 to 1935*. (Unpublished doctoral dissertation, University of Oxford)
21. W. Glade, *The Latin American economies*, p. 440
22. B. Roberts, *Cities of Peasants*, p. 71

23. C. Geertz, 1963, *Agricultural involution: the process of ecological change in Indonesia.* (University of California Press, Berkeley)
24. M. Santos, 1979, *The shared space: the two circuits of the urban economy in underdeveloped countries.* (Methuen, London)
25. B. Roberts, *Cities of Peasants*
26. R. Bromley (ed.), 1979, *The urban informal sector.* (Pergamon, Oxford)
27. B. Roberts, *Cities of Peasants*
28. H. Brookfield, 1975, *Interdependent development.* (Methuen, London)
29. M. Santos, *The shared space*
30. R. Bromley (ed.), *The urban informal sector;* M. Santos, *The shared space*
31. R. N. Gwynne, 1980, *The "biblioteca" industry of Caracas,* mimeo. (University of Birmingham)
32. A. S. Morris, 1979, *South America.* (Hodder & Stoughton)
33. S. Watanabe, 1970, Entrepreneurship in small enterprises in Japanese manufacturing, *International Labour Review,* 102, 6
34. H. Schmitz, 1982, *Manufacturing in the Backyard.* (Francis Pinter, London)
35. Ibid., pp. 155–6
36. International Labour Office (ILO), 1972, *Employment, Incomes and Equality: A Strategy for Increasing Productive Employment in Kenya.* (Geneva)
37. H. Schmitz, *Manufacturing in the Backyard,* pp. 161–2
38. L. R. Peattie, 1979, What is to be done with the "Informal Sector"? A Case Study of Shoe Manufacturers in Columbia, mimeo. (Massachusetts Institute of Technology)
39. G. Kitching, 1982, *Development and Underdevelopment in Historical Perspective.* (Methuen, London)
40. H. Brookfield, *Interdependent Development*
41. A. Quijano, 1974, The marginal pole of the economy and the marginalized labor force, *Economy and Society,* 3, 4, November, pp. 393–428
42. Ibid., p. 41
43. Ibid., pp. 410–11
44. El Mercurio, Santiago, Chile, 8 December 1982
45. B. Roberts, *Cities of Peasants,* p. 164
46. Ibid.
47. Ibid.
48. L. Kowarick, 1975, *Capitalismo e marginalidade na América Latina*
49. V. E. Faria, 1976, *Occupational marginality, employment and poverty in Brazil.* (Unpublished doctoral dissertation, Harvard University)

PART FOUR

INDUSTRIALISATION AND REGIONAL DEVELOPMENT

10 THEORETICAL PERSPECTIVES ON INDUSTRIALISATION AND REGIONAL DEVELOPMENT

This book has argued that industrialisation has been the single most important process of economic development in Latin America in the last forty years. Furthermore, it has been demonstrated that industrialisation is intimately linked to the process of urban concentration. As a result, many have argued that industrialisation is bound up with a distinctive spatial tendency in Latin America, that of increasing regional disparities. This is the phenomenon that signifies that in relative terms the more prosperous regions are enjoying higher growth rates than poorer regions so that the disparities between the more prosperous and the poorer regions are increasing rather than decreasing.

Regional disparities in Latin America have come to be associated with three broad patterns of spatial concentration and inequality. First, there are the large differences between the urban and rural areas. This can reflect the relative strengths of industry as against agriculture. In some areas, with both agriculture and industry dynamic or alternatively stagnant, disparities can be small e.g. São Paulo state (both dynamic) or Highland Bolivia (both stagnant). The biggest contrasts are where dynamic urban industry can be compared with stagnant rural agriculture; ironically, this can often be best seen in areas which have 'benefited' from industrial decentralisation programmes and most notably the Northeast of Brazil.

Secondly there are the economic and social disparities between the different regions of a country. This is a notoriously difficult index to measure and has been the basis of many statistical studies.[1] If a country has been divided into a small number of regions, disparities are statistically likely to be higher than if the country has been divided into a large number of regions. The scale of analysis is also important. When Goodman and Gilbert analysed regional disparities in the whole of Brazil, they quoted that disparities were increasing.[2] When Townroe and Keen confined their scale of analysis to the large state of São Paulo they found the reverse — that disparities were declining.[3] A third pattern of regional disparity has already been referred to in Chapter Six. It is the regional disparity caused by urban primacy when one city dominates the national urban structure. Most of the work on regional inequality in Latin America has concentrated on the latter two patterns of regional disparity.

The interest caused by regional disparities has largely been in terms of their relationship to the wider process of economic development. Williamson's paper on the relationship between regional disparities and economic development produced an empirical background to the subject.[4] Williamson's research was based on data taken from 24 countries of which only six could be categorised as less developed. However, from this research, he found that the nations with the largest regional differentials were drawn from a group of countries with intermediate levels of *per capita* income. Highly developed nations and those countries which had experienced only limited economic growth exhibited relatively small regional disparities. Williamson also analysed historical data on seven countries, and argued from their experience that 'increasing regional inequality is generated during the early development stages, while mature growth has produced regional convergence and a reduction in differentials'.[5] The diagrammatic representation of Figure 10.1 demonstrate the concept more clearly. As economic development begins to increase from a low level, relative disparities begin to widen. This period of divergence is followed by a period of convergence after a mature economy has been achieved.

Williamson very much concentrated on describing the patterns of regional disparity and did not investigate the processes behind the patterns. But there were attempts at explaining the reasons for increasing regional disparity in the early to middle stages of economic development. In the 1950s, 1960s and 1970s, it was generally agreed that Latin American countries were experiencing increasing regional disparities.

The most overtly industrial explanation for increasing disparities was that of Hirschman.[6] Writing in the 1950s, Hirschman argued for 'unbalanced' growth. Put very simply, 'unbalanced' growth recommended the concentration of state and private investment in a limited number of industrial sectors. However, growth generated by these sectors would create increasing demands and stimulate growth in other sections of the economy through processes of backward and forward linkage. But policies of unbalanced sectoral growth have similar spatial characteristics. Growth will be geographically concentrated around the location of the industries in which

Figure 10.1: Williamson's Model of the Relationship Between Economic Development and Regional Disparities Through Time

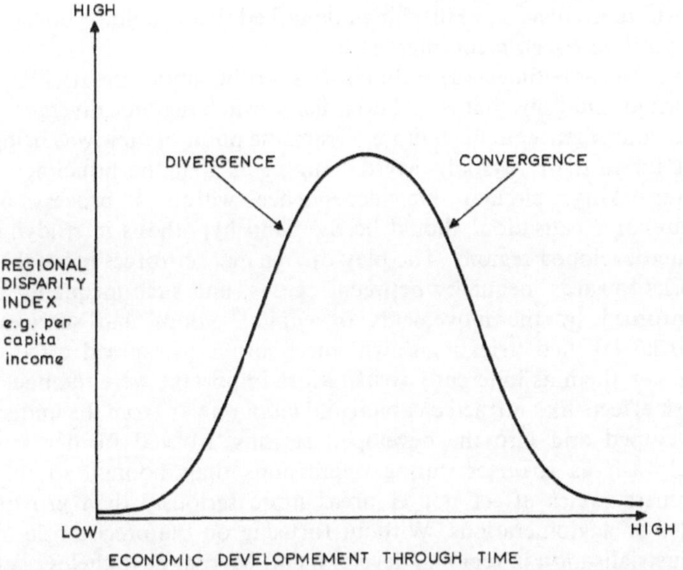

Source: J. G. Williamson, Regional Inequality and the Process of National Development: a description of the patterns, *Economic Development and Cultural Change*, 1965, 13, 3–45

governments decided to invest. Indeed these locations will form a type of growth pole (see Chapter Seven). Entrepreneurs in related sectors will concentrate at the location of the new industries. Growth in the developing region — 'North' in Hirschman's terms — will be paralleled by stagnation elsewhere (in the 'South') as skilled labour is withdrawn and savings are re-invested in the growing North.[7] Terms of trade will also go against the southern producer as it concentrates on exporting primary products to the 'North' in return for manufactured goods (see Chapter One).

Spatially-concentrated industrialisation in a dual-region model thus provides the basis for regional divergence in middle income countries according to Hirschman. He did, however, point to mechanisms by which the reversal of the process takes place — i.e. from regional divergence to convergence. Sufficient labour may leave the South for the industrialising North to increase the marginal productivity of labour in the South and raise *per capita* consumption. Furthermore, the North may increase its purchases from the South and its investments there. The force of these 'trickling-down' mechanisms as Hirschman described them could be boosted by positive government intervention.

At the same time as Hirschman was writing about the trickling-down mechanisms that would eventually switch regional divergence into convergence, a much more pessimistic point of view was being put forward by Myrdal.[8] Myrdal suggested that the principle of 'interlocking, circular inter-dependence within a process of cumulative causation' should be the main hypothesis in studying underdeveloped regions. The play of free market forces inevitably works towards inequality between regions, and such inequality is reinforced by the movements of capital, goods and services. Myrdal labelled 'trickling-down' mechanisms as 'spread effects' but saw them as inherently weak. More important were the backwash effects that attracted capital and labour away from the underdeveloped and into the developed regions. Myrdal further saw backwash as stronger during depressions than booms, so that business cycles affect poorer areas more seriously than growth poles or agglomerations. Without focusing on the precise role of industrialisation in regional development, Myrdal nevertheless saw the process of regional divergence continuing; regional convergence seemed distinctly unlikely.

In the 1960s, an institution that was influential in the regional planning process of Latin America was the Massachusetts Institute of Technology (MIT). In particular, the work of John Friedmann became influential. The MIT planning school saw regional convergence as difficult unless positive government intervention and planning took place:

> Although the government can make use of technological possibilities and changing demand structures in order to promote the development of peripheral economies, this is clearly a matter of deliberate planning. On the whole, unrestrained forces of a dynamic market economy appear to be working against a convergence of the center and the periphery.[9]

As with Myrdal, and despite being consultant to the Ciudad Guayana industrial project, Friedmann did not specifically examine the links between industrialisation and regional divergence. However, unlike Myrdal, he identified a number of mechanisms that were acting against regional convergence. Some of these mechanisms seem particularly powerful when they are more specifically related to manufacturing industry.

First, Friedmann identified a failure of diminishing returns to set in at the central city. In this way, entrepreneurs continued to benefit from economies of scale despite the creation of very large cities in which it could be assumed that the diseconomies of scale should begin to operate. In Chapter Seven, it was noted that the cost of labour and capital for manufacturers does not necessarily rise in the large city compared with smaller towns. Meanwhile, entrepreneurs benefit from a wide range of urbanisation economies.

Friedmann pointed to a second factor that has subsequently become much more significant — the export demand for goods produced at the centre. In the last twenty years, the most dramatic export growth has been recorded in the manufacturing sector (see Chapter Two), and as Chapter Six demonstrated most manufacturing activity now takes place in the central city of each country. As Latin American economies have become increasingly constrained by external forces in the 1980s (see Chapter One), the location of export growth becomes one of the major factors determining the spatial incidence of development. Location of export growth at the centre can only increase the contrasts between the prosperous and poorer regions.

Friedmann also noted that the centre was the major market for goods produced in the modern sector. As Chapter Six demonstrated the primate city is also the major producer of nationally-produced goods in the modern sector. This means that the primate city contains the majority of rapidly-growing manufacturing in the high technology sectors. Further, in terms of Friedmann's wider terms of reference, the early diffusion and production of modern goods to the centre again concentrate new innovations and associated growth impulses there as opposed to the periphery of a Latin American country.

The spatial framework within which Friedmann analysed regional divergence and convergence was that of two regions — centre and periphery, similar to that of Hirschman's North and South. Within this context, Friedmann saw entrepreneurs as failing to perceive investment opportunities in the periphery due to the distortion of their space preferences towards location in the centre. People in general and entrepreneurs in particular benefited from the intensity of culture contact at the centre and the concentration of high-order services there. Meanwhile, Friedmann identified the periphery as being unable to make the adjustments appropriate to constant change at the centre. This was because of high replacement rates due to population growth, the disruptive effects of emigration, a lack of capital, and a general national inability to see the regional problem from a nationwide perspective.[10]

Hirschman, Myrdal and Friedmann saw regional divergence as the major pattern of regional differentation in Latin America in the 1950s and 1960s. Reviewing the relationship between regional income disparities and economic development in the mid-1970s, Gilbert and Goodman concluded that:

> Williamson's proposition that regional incomes first diverge and then converge as per capita incomes rise does not meet with strong support from the available evidence. Recourse to more recent data for other less-developed nations to supplement his international cross-section, do not provide support either for convergence or for divergence in regional incomes.[11]

Furthermore, Gilbert and Goodman agreed with Hirschman and Friedmann that regional disparities are unlikely to diminish unless national governments adopt strong regional development programmes.

However, in the 1980s, a very different empirical and theoretical view has been put forward by regional economists working for and advising the World Bank.[12] The central concept that has been put forward is that of polarisation reversal.[13] Referring back to Figure 10.1, polarisation reversal is essentially the point where regional divergence becomes transformed into regional convergence. According to the definition used by the World Bank it is 'the point at which the growth rate of the secondary cities located outside the core comes to exceed that of the primate metropolitan centre'.[14] According to Townroe, this definition of polarisation reversal allows a continuing urban concentration in absolute terms where the population of the primate city is greater than that of the secondary cities combined, and so has the virtue of distinguishing the process of polarisation from the state of primacy.[15]

Research on polarisation reversal does not specifically refer to the process of industrialisation in regional development. Rather it looks at the general economic processes that have caused increasing urban concentration in Latin America (many of which were analysed by Friedmann) and attempts to look for the reversal of these processes. In so doing it is hoped to identify the turning-point between regional divergence and convergence.

> This turning point may be brought about by increases or decreases in the intensity of certain social and economic forces initially promoting concentration. At a given level, the interaction between these forces and the forces promoting deconcentration is such as to move the configuration of changes in the urban system from one direction to another. Alternatively, new forces enter the system. The underlying issue is that of changes in the relative advantages of different locations for the performance of economic activity. Indeed there is a close link here to the discussions on the role and evolution of secondary city growth centres.[16]

The analysis of polarisation reversal rests on two sets of factors. One set of factors is purportedly making the primate city less attractive for continued economic and demographic expansion. The other set of factors operates in the hinterland to promote the in-migration of firms and employment.

The factors that make the primate city less attractive are those that have had considerable impact in restricting the growth of large cities in the developed world but have had little impact in Latin America. Increases in congestion, crime and pollution in the large city do not seem to curb the immigration of people or the expansion of firms in the large Latin American city. Indeed many large cities welcome polluting firms if they substantially increase employment. Similarly, high costs of infrastucture do not have a repelling force mainly because the national government rarely passes on the high costs of infrastructure to the industrial or private consumer (see Chapter Six).

Pressures in the labour market of the large city are also put forward as stimuli to polarisation reversal.

> If metropolitan growth results in residential environmental deterioration and in higher taxes and longer journeys to work, then would-be immigrants are deterred and existing residents become prospective outmigrants, unless compensation is available in the form of high rewards from employment.[17]

Unfortunately in view of the brief review of labour markets in Chapter Nine, such argument seems inappropriate and distinctly 'Eurocentric'. On a statistical level, it was noted in Chapter Six that labour costs in the primate city were no greater, and often less, than in secondary centres. This is intimately linked to the massive immigration of labour from rural areas and smaller towns into the primate city. This causes what economists term in their impersonal way 'the bidding price of labour' to be held down. In a less technical way, it maintains low wages for those in work and high rates of unemployment and underemployment; the latter boost the number of the 'informal' and 'marginalised' sectors that were reviewed in Chapter Nine. According to World Bank advisers,[18] São Paulo presents a case of polarisation reversal. But according to other observers, the labour market argument cannot be used here. Storper concluded that since 1964 both the minimum wage and mean industrial wages had fallen in São Paulo.[19] The idea of employers in the large city having to increase wages in order to attract labour is not borne out even in the most successful manufacturing city in Latin America. Even in the relatively high-wage automobile sector of São Paulo, Humphrey has pointed out that high wages are only enjoyed by a small percentage of the total work force.[20]

One factor mentioned by the adherents of polarisation reversal does have an impact on repelling activity in certain cities — high and rapidly rising land prices. As was noted in Chapter Seven, land extensive activities, such as automobile manufacturing, do tend to migrate out of the large city to adjacent areas where land is cheaper.

The arguments for polarisation reversal, therefore, seem to rest on the validity of the second set of factors — those of a secondary city location. The argument refers to the possible behaviour of residents of the large city, of company executives and of immigrants, and is succinctly put forward by Townroe and Keen:

> The possibility of an alternative location for the resident will normally mean the guarantee or prospect of employment in a secondary city, access to public services of some national standard, and not more than a small fall in the real value of his money income. Similarly, for the company, outward movement from the metropolis will normally only be regarded as possible if the centres for deconcentration have relevant infrastructures,

> services and communications networks. For the resident, the migrant, the existing company or the new investment, location, or relocation to a secondary city therefore rests on prerequisites before a calculation of balance of advantage is undertaken.
>
> Prerequisites are not absolutes in these forms of locational behaviour, but they form a base upon which locational choices are made by trading off the relative advantages of the central metropolis against those of one or more of the secondary cities. If the information is available, a company for example will compare relative land values, transport costs and wage levels just as the resident will compare relative house prices, journey to work costs and income levels. Changes in the aggregate pattern of choices will induce polarization reversal.[21]

Townroe and Keen's statistical analysis of the São Paulo city region seems to demonstrate that polarisation reversal took place in the 1970–80 decade due to rapid population expansion of secondary cities rather than decline in Metropolitan São Paulo itself. Between 1970 and 1980, the population of Metropolitan São Paulo increased at the high annual rate of 4.6 per cent. But the secondary cities of São Paulo state with a population of over 100,000 grew at an average annual rate in excess of 5.3 per cent (see Table 10.1).

However, the argument for polarisation reversal and the empirical interpretation of it taking place in São Paulo state are highly dubious. Why was the analysis of polarisation reversal confined to the state of São Paulo and not examined at a wider geographical scale — the Southeast Region or Brazil itself? When Gilbert and Goodman analysed regional disparities in Brazil as a whole[22] in the mid-1970s, they found evidence for continued regional divergence. Dickenson[23] noted that the richest region of Brazil (Guanabara) had a regional *per capita* income nine times greater than the poorest region (Maranhão).

An equally tenable interpretation of the data in Table 10.1 is that it reflects the diffusion of population growth by distance away from a highly dynamic national (not regional) core. The growth of the secondary cities in São Paulo state is thus a result of overspill (as outlined in Chapter Seven) rather than autonomous expansion. This interpretation is reinforced by the location of the high-growth secondary cities. Of those cities located within a radius of 150 kilometres from Metropolitan São Paulo, three out of four grew faster than the metropolitan area between 1970 and 1980. Only one out of

Table 10.1: São Paulo State: Comparison of Average Annual Population Growth Rates

	1950–60	1960–70	1970–80
Metropolitan São Paulo	6.8	5.8	4.6
Secondary cities			
20,000– 50,000 (n=45)	5.6	4.0	4.1
50,000–100,000 (n=14)	6.2	4.9	4.3
100,000–250,000 (n=9)	6.0	5.2	5.3
250,000+ (n=2)	5.5	5.3	5.4

Source: P. M. Townroe and D. Keen, Polarization reversal in the state of São Paulo, Brazil, *Regional Studies*, 1984, 18, 1, 45–54

five cities located beyond 150 kilometres grew faster than Metropolitan São Paulo. Such a spatial distribution of urban growth seems to reinforce a hypothesis of 'strengthening centralisation' rather than 'polarisation reversal'.

I have dwelt on the concept of polarisation reversal because it has become popular during the 1980s due to its links with World Bank funding. Unfortunately one is left with the impression that much of the work on polarisation reversal in Latin America has been carried out by specialists on regional problems of the developed world who have taken their armoury of concepts and methods with them to study a very different process of urban and regional development — and not modified or adapted them to the new circumstances. Furthermore, one has the impression that data are collected from specific areas to back up the concept. However, as in the case of São Paulo state, the selection of the spatial unit to be studied has little theoretical justification. The processes of regional divergence and convergence operate at a national level not at the level of a city region. Although São Paulo state has a population larger than most Latin American countries (25 million), it is operating and functioning within a much larger national context. One cannot understand the dynamic functioning of São Paulo state without referring to the large quantities of resources consumed there that come from other Brazilian regions (steel, petrochemicals, aluminium, beef and wheat to name but a few) or the large numbers of immigrants attracted from other regions and most notably Brazil's populous Northeast. Within this wider national context of Brazilian growth, the rapid growth of secondary cities less than 150 kilometres from Metropolitan São Paulo is surely a case of the centralised metropolis spreading its growth outwards, — not a dramatic case of polarisation reversal as has been suggested by World Bank economists.

In the more general terms of political economy, the concept of polarisation reversal represents the empirical search for justifying the idea of a relatively smooth path of regional development within the overall framework of a capitalist model of growth. This can best be understood by referring to Alonso's 1980 article on the five bell shapes of development.[25] They refer to stages of economic development, social inequality, regional inequality, geographic concentration and demographic transition. According to Alonso, the explanation for the bell shape of development (see Figure 10.1) is the characteristic of unbalanced growth in the early stages of development, followed by progressive social, economic and geographic integration in the later stages. 'Their paths may be deflected by the concrete circumstances of history, including environmental conditions, exogenous shocks, and purposeful national policy.'[26] But, generally speaking, all countries follow these paths, and furthermore there is considerable interaction between the five phenomena. In terms of regional development and the right-hand tail of the curves of regional equalisation and concentration, considerable emphasis is given to the US experience.

> Regional equalization of *per capita* incomes in the United States has proceeded rapidly among the nations' geographic regions and between metropolitan and non-metropolitan areas. It is even possible that the *per capita* income curves of the Sunbelt and the Frostbelt will not stop at convergence, but may in fact cross over.
>
> Regional concentration has reversed. Not only are the majority of the large U.S. metropolitan areas declining absolutely in population but substantial net migration out of metropolitan into nonmetropolitan areas has characterized the scene since about 1970.[27]

Regional inequality and regional concentration are thus temporary phenomena that with continued economic development, social and demographic change will gradually decline.

Can the development paths of developed countries be extrapolated to portray the future of less developed countries in general and Latin American countries in particular? A rapidly-increasing body of regional development literature states that they cannot. The principal argument concerns the fact that while the USA is very much the central motor of modern world capitalism, most of Latin America plays a peripheral or even dependent role in world capitalism. As a result processes that occur in the USA need not occur at a later date in other parts of the world. In terms of regional development in Latin America, the contradiction between territory and function has become the centre of much debate.

The most influential person to enter this debate has been John Friedmann, in a very different guise and with radically changed arguments to those of the 1960s. In the 1960s, Friedmann had argued (along with other urban planners at MIT) that city growth accompanied modern development in less developed countries. In particular, he argued that urbanisation was significant in a country's passage from a traditional to an innovative society, from a limited culture to one of constantly expanding opportunities, and from an elitist to a mass political system 'striving for national integration'.[28] In this way, Friedmann and the MIT school had favoured a growth centre strategy for regional development. Large cities were more effective than small towns in promoting a more innovative society and consequently resources should be concentrated in the former. The primate city itself would be a centre of great innovation. Thus, it would be necessary to create strong regional growth centres in the periphery of each Latin American country so as to assist the downward percolation of development impulses through the urban hierarchy from the primate city to small towns and their surrounding rural areas. Figure 10.2 shows the Friedmann-inspired regional development strategy of Chile's Christian Democrat government (from 1964 to 1970).[29] Outside the core of Santiago, twelve regional growth centres were created. In these regional cities, public and private resources were concentrated so as to maximise the trickle-down effect of innovation and modernisation to the small towns and rural areas of Chile.

By 1979, however, Friedmann was to denounce vehemently the regional planning policies that he had previously favoured.

> With the growth centre doctrine as its principal tool, spatial development planning became the handmaiden of transnational capital.
>
> The growth centre doctrine is completely attuned to the ideology and planning approaches of transnational corporations.[30]

Figure 10.2: The Regional Development Strategy of Chile's Christian Democrat Government, 1964–70

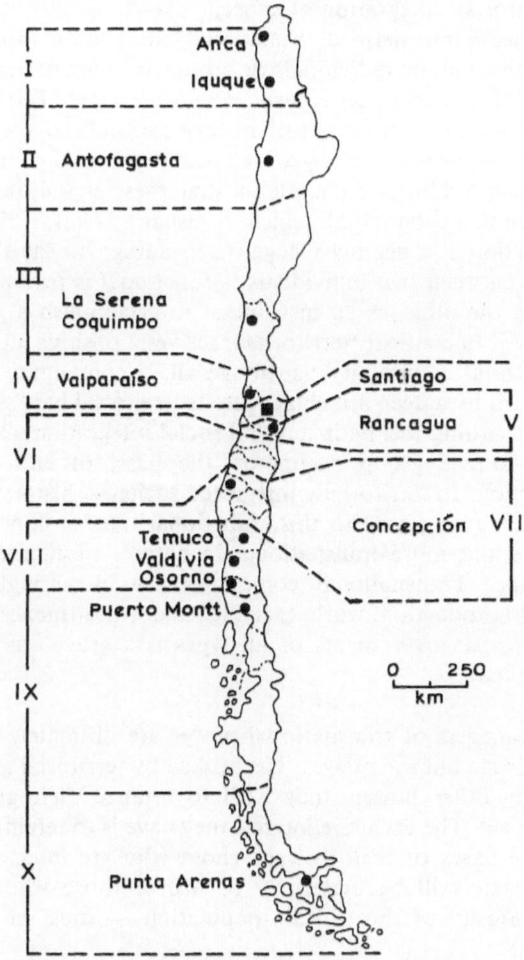

Note: Names = regional growth centres; numbers = 12 regions outside the metropolitan area of Santiago
Source: ODEPLAN

Such a dramatic revision of a theory and policy that he had extensively and enthusiastically put forward less than ten years earlier was due to a perceived contradiction between the 'functional' and 'territorial' integration of space.

Friedmann defined 'functional integration' as referring to linkages among individuals or groups who are organised into hierarchical networks on a basis of self-interest.[31] 'Territorial integration' refers to those ties of history and collective experience that bind the members of a geographically bounded community to one another.[32] Hilhorst complains that these definitions are given a meaning beyond that which is usual.[33] Thus, Friedmann gives 'functional' a distinctly negative meaning: 'to say that a relationship between two individuals is functional is to imply that one is using the other as an instrument to accomplish a purpose of his own'.[34] In contrast 'territorial' receives a positive interpretation: 'a territorial community, is, above all, a community of sentiment, formed by a deep attachment to its territorial base'.[35]

According to Friedmann 'all social integration above the small, face-to-face group occurs on the basis of either territory or function. In territorially integrated societies, history and place are one'.[36] Juxtaposed to this, functional relationships have become more and more transnational in nature, particularly in the last decade. Transnational companies are increasingly dominating world production, world exchange and world finance. The threat to territorial governments of all types is a grave one, according to Friedmann.

> Managers of transnational power are ultimately responsible to no one but themselves. Unchecked by territorial and, indeed, by any other power, they wish to totalize their grasp upon the world. The exclusive interest they have is to retain and to enlarge the bases of their power. Those who are integrated into their system will be materially rewarded; those who are not — a majority of the world's population — must be controlled by force.
>
> Transnational development occurs according to the principles of an exchange economy in which efficiency in production and equity in distribution are structually separated in theory as well as fact. Given the global approach of the transnationals, this implies that a majority of the world's population is destined to remain outside the system of transnational power and, therefore,

> poor, exploited, and coerced. But even those who are within the system will be exploited to serve the transnational managers whose power to decide is absolute and knows no interest beyond its further aggrandizement.[37]

The managers of transnational power operate in the world economy through the urban system. Large cities, including the large primate cities of Latin America, lend themselves to such new international roles. The functions of the worldwide corporations are concentrated here. Corporate influences in individual countries permeate down the urban hierarchy away from the primate city. It is in this sense, according to Friedmann, that the growth centre doctrine became the handmaiden of transnational capital.

Friedmann's new position is not, itself, new. The potential conflict between international capital and individual nations has been a critical area of study for neo-Marxist writers ever since Lenin. Indeed a clearer framework within which to analyse the implications of the growth of multinational corporations on less developed countries was put forward by Hymer.[38] Hymer distinguished three levels of business administration in the multinational corporation. Level Three, the lowest level, is concerned with managing the day-to-day operations of the enterprise. Level Two, which first made its appearance with the separation of head office from field office, is responsible for coordinating the managers at Level Three. The functions of Level One, top management, are goal determination and planning. This level sets the framework in which the lower levels operate. In the family firm, all three levels are embodied in the single entrepreneur. In the national corporation, a partial differentiation is made in which the top two levels are separated from the bottom one. In the multinational corporation, the differentiation is far more complete. Level One is completely split off from Level Two and concentrated in a general office whose specific function is to plan strategy rather than tactics.

In terms of location, Level Three activities tend to spread themselves over the globe according to the pull of manpower, markets and raw materials. As we noted in Chapter Six, there is a strong tendency for the location of plants in Latin America to be drawn to the core area of that country — due to the transport network, better infrastructure, easier contact with government, more reliable linkage with suppliers and a general desire to reduce risk and uncertainty. Some plants are drawn to the periphery of Latin American countries normally for resource considerations. If the desire to reduce risk and uncertainty can affect the location of Level Three activities within countries, it would appear to exert an even stronger force on the location of Level Two activities. Such activities tend to be heavily concentrated within the lesser world metropolitan areas; in Latin America, these constitute the major metropolitan area of each country — São Paulo, Buenos Aires, Mexico City, Santiago and so on. Location is governed by the need to be near other corporations and good communication systems. Meanwhile, Level One activities, the strategic decision makers, are generally located in the major world metropolitan areas, depending on the predominant nationality of the multinational corporation — New York, London, Paris and Tokyo for example.

Hymer's analysis revolved around manufacturing multinationals. Cohen more recently argued that multinational enterprise is becoming more diverse and venturing strongly into finance and services. As a result, changes in the corporation and in the structure of the advanced corporate services have led to the 'emergence of a series of global cities which serve as international centres for business decision-making and corporate strategy formulation'.[39] Multinational corporations are viewed in a similarly negative way to that put forward by Friedmann and the contradiction between territory and function is again central to the argument.

> As international boundaries become blurred by the increasingly global nature of corporations, numerous contradictions will arise within the world hierarchy of cities. First, there will be contradictions that will arise because private institutions, particularly large multinational corporations and banks, are able to undermine or contravene established government policy. This contributes to the erosion of the position of certain traditional centres of government policy where corporate head offices or major financial institutions are not present in large numbers.[40]

If one looks more closely at the actual process of industrialisation in Latin America the theoretical framework of conflict between territory and function (or between territorial and multinational organisations) has numerous problems. First of all, the blanket assumption the multinational enterprise is inimical to all meaningful development is not borne out by the evidence. Evidence of improper behaviour by multinational enterprise in Latin America does exist. On the other hand, as was demonstrated in Chapter Three, multinational enterprise has brought new technologies, new management techniques, new financial systems and has assisted in key policies such as export promotion. Multinational enterprises are neither spectre nor spearhead, although aspects of both can be found.

Secondly, the neo-Marxist idea[41] of multinational enterprise having as its major aim the steady extraction of the economic surplus of less developed countries (and particualrly the surplus of the periphery of each country) in order to concentrate resources at the world's centre (see Chapter One) has had less empirical foundation since the Second World War. As Hamilton forcefully argues the key features of the global pattern of multinational investment since 1945 have been the rising concentration of activity in the developed countries and the sharp relative decline in importance of investments in the developing countries.[42] For example, multinational enterprise now accounts for nearly 40 per cent of economic activity in Western Europe — before 1939 it was responsible for only 7 per cent.[43] Latin America now only accounts for 15 per cent of global investment from multinational companies.

One reason for this is the nationalisation of multinational interests in many Latin American and other less developed countries. This has been particularly evident in activities linked to resource extraction and refining. Oil production, for example, used to be controlled by seven multinational companies known as the 'Seven Sisters'. Now they control less than one quarter of world production. In Latin America, multinational companies specialising in copper, tin, iron ore, aluminium and sugar have similarly been nationalised.[44] State companies such as PETROVEN, PEMEX and CODELCO have replaced them (see Table 10.2).

Another development that considerably complicates the neo-Marxist framework of conflict between territorial and multinational organisations is the rise of the Third World multinational company. Table 10.2 lists thirty Latin American companies that had sales greater than one billion dollars in 1982. Most of these companies are multinational. For example, the company ranked seventeenth, the Pão de Açucar group constitutes the major supermarket chain of Brazil. Nevertheless, it also has forty stores in Portugal, one in Angola and an international subsidiary in Luxembourg. Most large Latin American companies have similar

Table 10.2: Latin American Companies with 1982 Sales Greater than one Billion Dollars

Rank	Company	Country	Sales (US$ millions)	Nature of Business
1	PETROBRAS	Brazil	17,091	Oil
2	PETROVEN	Venezuela	16,482	Oil
3	PEMEX	Mexico	10,020	Oil
4	YPF	Argentina	4,455	Oil
5	SIDERMEX	Mexico	3,410	Steel
6	Colombia Petróleos	Colombia	2,152	Oil
7	TELEBRAS	Brazil	2,070	Telecommunications
8	Gas del Estado	Argentina	1,927	Gas
9	SEGBA	Argentina	1,907	Electricity
10	CONASUPO	Mexico	1,688	Food
11	CODELCO	Chile	1,660	Copper
12	CEPE	Ecuador	1,529	Oil
13	Cerveceria Popular	Venezuela	1,484	Beer
14	Valores Industrial	Mexico	1,482	Food
15	SIDERBRAS	Brazil	1,306	Steel
16	Telefonos de Mexico	Mexico	1,302	Telecommunications
17	Gp. Pão de Açucar	Brazil	1,295	Supermarkets
18	Agua y Energia	Argentina	1,260	Electricity
19	ENTEL	Argentina	1,257	Telecommunications
20	Desc. Soc. Fom. Ind.	Mexico	1,214	Chemicals
21	Cia. Vale do Rio Doce	Brazil	1,184	Iron Ore Mining
22	Ind. Votorantim	Brazil	1,183	Mining
23	Aerolineas Argentinas	Argentina	1,143	Airline
24	Gpo. Ind. Alfa	Mexico	1,124	Steel
25	COPEC	Chile	1,116	Fuel
26	Aurrera	Mexico	1,096	Commerce
27	Petróleos del Peru	Peru	1,064	Oil
28	Emp. Minera del Peru	Peru	1,061	Minerals
29	Eletropaulo	Brazil	1,052	Electricity
30	Petro Ipiranga	Brazil	1,032	Oil

Source: The South, July 1984

international interests, whether in supply, markets, investment or technology. Such a pattern destroys the Friedmann contention that Third World countries have no access to transnational power.

Indeed such is the complexity of the process of industrialisation in Latin America that transnational power is limited. As Chapter Four demonstrated, multinational enterprise constitutes but one of three institutions important in the industrialising process. State firms, large though small in number, and private national firms constitute the other two parts of the Triple Alliance. A more useful framework within which to analyse transnational power in Latin American countries is that of bargaining — bargaining between the multinational enterprise, on the one hand, and the territorial organisation (national or regional government), on the other. Some studies have shown how the host government can often hold the strongest bargaining position — as with the nationalisation of foreign oil interests in Venezuela.[45]

According to Friedmann's more recent analysis, transnational enterprise should be drawn to those countries where the multinational's bargaining power is strong and the host government's weak. But as Chapter Four demonstrated this has not been the case. Multinational enterprise has been primarily attracted to the larger countries of Latin America whose governments can exert considerable bargaining power on the terms of multinational entry. Multinational computer firms in Brazil, for example, had to agree to export a substantial proportion of their production in return for entry into the Brazilian market. Meanwhile, multinational enterprise has been much less evident in the smaller countries of Latin America, some of which have eliminated almost all restraints on transnational entry.

Friedmann's policy recommendation follows that of other neo-Marxist writers in favouring an end to transnational influence in Latin America and other less developed countries. At the same time, a shift from urban-based to rural-based development is recommended in what Friedmann terms an 'agropolitan approach'.[46] After fervently promoting the innovative and modernising influences of an urban-based approach in the 1960s, Friedmann fervently argues for the equalising benefits of rural-based development in the late 1970s. He argues that the Chinese model of rural-based development is the ideal to be followed.

However, straying briefly from the Latin American continent, it is worth putting into context the policy recommendations of Friedmann and other writers on Latin America[47] for rural-based regional development more closely aligned with the Chinese model. For development writers whose area of study is China rather than Latin America, such simplified ideas of the Chinese model are alarming. Chinese development in the last forty years, despite a powerful rural development programme, has relied heavily on a massive industrialisation programme. In 1974, the Chinese industrial sector was responsible for 54 per cent of Chinese GDP, one of the highest proportions in the world; the Chinese rural sector was responsible for only 20 per cent of national GDP.[48] Furthermore, in 1974, urban (as opposed to rural) industry accounted for 75 per cent of the gross value of industrial output, and 'large-scale urban plants in the metallurgy, machine building, automotive, aircraft, electronics and oil extraction industries accounted for the bulk of this'.[49]

The role of urban-based industrialisation in the Chinese model of development and its comparison with a country, Tanzania, that has followed the 'agropolitan' approach recommended by Friedmann is put into context by Kitching as follows:

> In China it was the 'capital-intensive' producer goods industries of Manchuria and the relatively capital-intensive engineering industries of Shanghai and the other coastal cities which provided the machinery for China's rural industrialization programme and which, even more importantly, provided both directly and indirectly the inputs (cement, fertilizer, iron and steel, agricultural implements and machinery) which made possible the continued expansion and intensification of China's agriculture. In addition, it was in those industries that an industrial proletariat and technical class were formed whose skills were then diffused into the rural areas. We have already noted how the effective absence of these 'motor' industries in Tanzania left the rural development effort there bereft of both inputs and expertise.[50]

The 'agropolitan' approach favoured by Friedmann and other writers such as Slater, would appear inappropriate to Latin American conditions on two grounds. First, there is little idea of the crucial role that industrialisation plays in the process of economic development. The argument for agropolitan development is furthermore based on a misinterpretation of the success of the Chinese model of development. Second, as Table 10.3 demonstrates, Latin America is already a highly urbanised continent. The seven largest countries of Latin America, Brazil, Mexico, Colombia, Argentina, Peru, Venezuela and Chile, all have more than 64 per cent of their population in urban areas. A rural-based regional development strategy would be anachronistic in such highly-urbanised societies.

Undoubtedly, the post-war process of industrialisation in Latin America has been characterised by a spatial concentration of

Table 10.3: Urban Population as a Percentage of Total Population, 1981

	Per Cent		Per Cent
Venezuela	84	Nicaragua	54
Uruguay	84	Ecuador	45
Argentina	83	Bolivia	45
Chile	81	Costa Rica	44
Brazil	68	El Salvador	41
Mexico	67	Paraguay	40
Peru	66	Guatemala	39
Colombia	64	Honduras	36
Panama	55		

Source: World Bank, *World Development Report, 1983* (Washington)

national industry with little historical precedent. As a result urban concentration has been substantially fuelled. Linked to both these phenomena has been the evolution of considerable spatial disparities in both regional income and welfare. Chapter Seven has already demonstrated that the decentralisation of industry is not occurring in any significant way at the present time. The major process of decentralisation that is observable is an onion-like growth out from the central metropolitan core itself — made up of large numbers of short-distance moves of firms already established there. Despite certain commentators claiming to have found polarisation reversal, there is little evidence of this occurring on any national scale. The spatial concentration of industrialisation thus continues and the already large disparities in regional income and welfare do not seem to be declining. It would however be inappropriate at the present highly urbanised state attained by most Latin American countries for their governments to restrict industrial growth and favour a radically different rural-based approach. As Kitching has demonstrated, when governments have adopted these latter policies, economic stagnation rather than growth results. All successful attempts at economic development have benefited from the dynamic impulses provided by industrialisation. At present, Latin America is developing a radically different spatial system to any other continent as it industrialises. However the net benefits of such a system in terms of social and economic development in the widest sense would still appear greater than the net benefits of an alternative system.

References

1. P. M. Townroe and D. Keen, 1984, Polarization Reversal in the State of São Paulo, Brazil, *Regional Studies*, 18, 1, pp. 45–54
2. A. Gilbert and D. Goodman, 1976, Regional income disparities and economic development: a critique, in A. Gilbert (ed.), *Development planning and spatial structure*. (John Wiley, London)
3. P. M. Townroe and D. Keen, Polarization Reversal, *Regional Studies*
4. J. G. Williamson, 1965, Regional inequality and the process of national development: a description of the patterns, *Economic Development and Cultural Change*, 13, pp. 3–45
5. Ibid.
6. A. O. Hirschman, 1958, *The strategy of economic development*. (New Haven, Conn.)
7. Ibid.
8. G. Myrdal, 1957, *Economic theory and underdeveloped regions*. (Duckworth, London)
9. J. Friedmann, 1966, *Regional development policy: a case study of Venezuela*. (MIT Press, Cambridge, Mass.), p. 18
10. Ibid.
11. A. Gilbert and D. Goodman, Regional income disparities
12. J. V. Henderson, 1980, A framework for international comparisons of systems of cities, *Urban and Regional Report* No. 80–3, The World Bank, Washington D.C. D. Keen, 1982, The relationship between urban infrastructure and industrial development among cities of São Paulo State, *National Spatial Policies* Working Paper No. 12, The World Bank, Washington D.C. J. F. Linn, 1979, Urbanisation trends, polarization reversal and spatial policy in Colombia, *Urban and Regional Report* No. 79–15, The World Bank, Washington D.C.
13. H. W. Richardson, 1980, Polarization reversal in developing countries, *Papers of the Regional Science Association*, 45, pp. 67–85
14. H. W. Richardson, 1977, City size and national spatial strategies in developing countries, *World Bank Staff Working Paper, No. 252. The World Bank,* Washington D.C.
15. P. M. Townroe and D. Keen, 1984, Polarization Reversal, *Regional Studies*, 18, 1, p. 46
16. Ibid.
17. Ibid.
18. Ibid.
19. M. Storper, 1984, Who benefits from industrial decentralization? Social power in the labour market, income distribution and spatial policy in Brazil, *Regional Studies*, 18, 2, pp. 143–64
20. J. Humphrey, 1982, *Capitalist control and workers' struggle in the Brazilian auto industry*. (Princeton University Press, Princeton)
21. P. M. Townroe and D. Keen, 1984, Polarization Reversal, *Regional Studies*
22. A Gilbert and D. Goodman, Regional income disparities
23. J. Dickenson, *Brazil*
24. P. M. Townroe and D. Keen, Polarization Reversal, *Regional Studies*
25. W. Alonso, 1980, Five bell shapes in development, *Papers of the Regional Science Association*, 45, pp. 5–16
26. Ibid.
27. Ibid.
28. J. Friedmann, 1969, The role of cities in national development, *American Behavioural Scientist*, 12, 5
29. Oficina de Planificación Nacional (ODEPLAN), 1968, *Politica de desarrollo,*

Santiago

30. J. Friedmann and C. Weaver, 1979, *Territory and Function.* (Edward Arnold, London) pp. 186–8

31. J. Friedmann, 1976, On the contradictions between city and countryside, (UCLA, mimeo), p. 11

32. Ibid.

33. J. G. M. Hilhorst, 1981, Territory vs Function: a new paradigm?, *Institute of Social Studies, The Hague, Occasional Papers*, No. 89

34. J. Friedmann, On the contradictions between city and countryside

35. Ibid.

36. Ibid.

37. J. Friedmann and C. Weaver, 1979, *Territory and Function*, p. 189

38. S. Hymer, 1972, The multinational corporation and the law of uneven development, in J. Bhagwati (ed.), *Economics and world order from the 1970s to the 1990s*. (Collier-Macmillan), pp. 113–40

39. R. B. Cohen, 1981, The new international division of labour, multinational corporations and urban hierarchy, in M. Dear and A. J. Scott, *Urbanization and Urban Planning in Capitalist Society*. (Methuen, London), pp. 287–315

40. Ibid., p. 308

41. A. G. Frank, 1971, *Capitalism and underdevelopment in Latin America*. (Penguin, Harmondsworth)

42. F. E. I. Hamilton, Multinational enterprise: spectre or spearhead, in D. Watts, D. Drakakis-Smith and C. Dixon, *Multinational Companies and the Third World*. (Croom Helm, London, forthcoming)

43. Ibid.

44. P. E. Sigmund, 1980, Multinationals in Latin America

45. Ibid.

46. J. Friedmann and C. Weaver, *Territory and Function*, p. 193

47. D. Slater, 1975, Underdevelopment and spatial inequality; approaches to the problems of regional planning in the Third World, *Progress in Planning*, 4, 2, pp. 97–167

48. G. Kitching, 1982, *Development and Underdevelopment in Historical Perspective*. (Methuen, London)

49. Ibid.

50. Ibid., p. 139

11 INDUSTRIALISATION AND REGIONAL DEVELOPMENT IN CHILE

So far in this book we have examined the process of industrialisation and its relationship with urban and regional development for the seventeen Spanish- and Portuguese-speaking countries of the Latin American mainland. Throughout we have been intent on extracting general principles and making general observations that apply to most of the countries under study. However, a certain depth is lost in such a task. In order to make some attempt to rectify this, this last chapter seeks to examine the process of industrialisation and its relationships with urban and regional development in just one country of Latin America, Chile. The Chilean case is an interesting one to examine because while being one of the smaller countries of Latin America (population of 11.3 million in 1981) it is also one of the most industrialised. In the twentieth century it has also undergone fundamental changes in industrial policy and direction. As a result, this case study will examine Chilean industrialisation and its relationship to the spatial development of the country in four stages:

1. Early attempts at industrialisation with a free-trade model (pre-1928).
2. Protectionism and government planning (1928–57).
3. Industrial stagnation and radical alternatives (1957–73).
4. A return to a free-trade model (1973–85).

Early Attempts at Industrialisation (Pre-1928)

Despite some isolated attempts at industrial protection before 1928, Chile's economic history before the 1928 Tariff Act was dominated by the theory and practice of free trade. As a result, Chile fell into the classic pattern of an exporter of primary goods (notably nitrates and copper) and an importer of manufactured goods. However, some significant industrial developments had taken place. Many of these were associated with export activities, such as the refining of nitrates and copper in North Chile or flour milling in the Central

Valley. Other industrial developments reflected local enterprise in processing resources for the national market. One example of this was the production of high-quality woollen and worsted cloths in the Tomé district near Concepción where by the 1870s 'local enterprise was supplying not only wool, but sulphuric acid, tartaric acid, sulphate of copper and soda ash.'[1] Other industrial developments reflected the impact of the friction of distance. Because Chile was so far from the industrialised countries of Europe and North America, there were considerable cost savings in producing heavy goods locally rather than importing them. One example of this was heavy engineering, where the influence of British-descended residents of Valparaíso was important. In 1898, the report of John F. Caples, the US consul in Valparaíso, gave evidence of a significant local engineering industry.

> There are two well-appointed steam factories in Valparaíso. One is owned by Messrs Balfour, Lyon & Co., and the other by Messrs Hardie & Co. The best equipped factory of the kind in Chile is at Caleta Abarca, about 4 miles from Valparaíso . . . [it] has constructed numbers of locomotives, many hundreds of cargo cars, and some passenger cars. With the exception of the wheels, many entire locomotives, including boilers, have been built by Messrs Lever, Murphy & Co.[2]

As in Brazil and Argentina, the role of immigrant entrepreneurs was important in the early industrial development of Chile. In 1928, first-generation immigrants accounted for over 23 per cent of all industrial entrepreneurs (see Table 11.1). Their influence was much greater in certain key provinces — in the northern mining provinces of Arica, Tarapacá and Antofagasta, in the extreme southern province of Magallanes where Yugoslav entrepreneurs were particularly important, and in the two major industrial provinces of the time, Valparaíso and Santiago. Indeed 58 per cent of all immigrant entrepreneurs had located in these two provinces according to the 1928 Industrial Census. The importance of immigrants as entrepreneurs can be gauged by comparing their proportional significance as entrepreneurs with that of the total industrial labour force. While accounting for 23 per cent of total entrepreneurs, immigrants only accounted for 5 per cent of the industrial labour force — even less in Valparaíso and Santiago.

Immigrant entrepreneurs thus helped to create an important

Table 11.1: Chile: Immigrants as Industrial Entrepreneurs (Owners, Partners, Lessees) and their Distribution by Province in 1928

Province	Immigrant industrial entrepreneurs	Total industrial entrepreneurs	Immigrants as percentage of total	Immigrants as percentage of total industrial labour force
Arica	26	76	34.2	14.4
Tarapacá	77	198	38.9	9.8
Antofagasta	79	267	29.6	8.9
Atacama	17	117	14.5	6.7
Coquimbo	14	273	5.1	5.0
Aconcagua	20	124	16.1	3.2
Valparaíso	321	1,190	27.0	4.5
Santiago	927	2,817	32.9	4.7
Colchagua/O'Higgins	37	438	8.4	2.9
Talca/Curicó	64	331	19.3	4.2
Linares/Maule	24	354	6.8	2.7
Ñuble	33	446	7.4	2.5
Concepción/Arauco	166	722	23.0	4.2
Bío-Bío/Malleco	59	386	15.3	4.1
Cautín	92	540	17.0	4.5
Valdivia/Osorno	66	547	12.1	4.1
Llanquihue	22	156	14.1	4.3
Chiloé	7	152	4.6	3.2
Magallanes	114	207	55.1	16.9
CHILE	2,165	9,343	23.2	4.9

Source: Dirección General de Estadística, *I Censo Nacional de Industrias (1928)* (Santiago, 1930)

industrial city, Santiago even before protection took place. By 1928 Santiago had 38,000 people employed in industry, 43 per cent of the Chilean total.[3] The export economy phase of Chilean development, that had built up in a distinctly cyclical way from 1840 to 1928, had reinforced a spatial pattern of population concentration (see Chapter Eight). In particular, the appropriation by the central government in Santiago of export taxes on the nitrates and copper from the northern provinces caused a concentration of government revenue and spending in the capital city, which in turn led to rapid growth in both employment and population. As a result, Santiago became the major consumer market of Chile and industrial entrepreneurs, both immigrant and national, were attracted there. By 1928, Santiago had already become nationally dominant in such market-oriented sectors as footwear, clothing, textiles, glass, furniture, paper and printing.

Three other major industrial centres had been formed in the early phase of industrialisation. One was Santiago's port, Valparaíso. Apart from its heavy engineering capacity, it had become renowned for a variety of industries linked to its port function — tobacco, sugar, chocolate, biscuit and chemical industries, in particular. Concepción, 500 kilometres to the south of Santiago, was able to develop a wide range of industry partly due to the high transport costs that Santiago's manufacturers had to pay to arrive in southern Chile. Although Concepción was the major administrative and service town of southern Chile, manufacturing expansion largely took place in a collection of neighbouring small towns. Tomé became the centre for woollen textiles, Chiguayante for cotton textiles, Penco for ceramics and sugar, Lirquén for glass, Lota for refractory bricks, Talcahuano for fish-processing and San Vicente for ship repairing.[4] Nearly four hundred kilometres to the south of Concepción, a third industrial centre, Valdivia, had grown up, strongly dependent on German immigrant entrepreneurs. Along with the industrial towns of Concepción, it competed for the market of the Chilean south, which had witnessed considerable colonisation in the last quarter of the nineteenth century, mainly by Germans. Valdivia's industrial importance lay in brewing (due to the German influence), leather, footwear, food, timber and furniture. Near Valdivia, at Corral, was Chile's only blast furnace, making Valdivia an important metal working centre. In serving the south Chilean market, it is interesting to note that the industrial specialisations of Concepción and Valdivia were complementary rather than competing. Between them, the three regional industrial centres of Chile generated 34 per cent of Chile's manufacturing employment in 1928. The free-trade model had then been associated with a reasonably decentralised pattern of industry.

Protectionism and Government Planning (1928–1957)

The free-trade model came to an abrupt end at the end of the 1920s when the world depression hit Chile particularly hard. The value of Chile's exports declined to one quarter of their former value in a matter of three years. The Chilean government already had passed a Tariff Act in 1928 and rapid use was made of it. Between 1928 and 1931, tariffs increased by an average of 71 per cent and affected 73 per cent of imports.[5] In 1933 and 1934, additional tariffs equivalent to 100 per cent were imposed. No general revision of tariffs took place after this, although in certain sectors further tariff increases took place. Thus, tariff protection of Chilean industry changed dramatically in less than five years. Before 1928, tariffs offered little protection (generally less than 25 per cent). Between 1928 and 1932, protection increased dramatically and after 1933, Chilean industry was excessively protected. One example of this excessive protection was the cost of a tyre in 1942. The c.i.f. cost of a tyre at the port of Valparaíso was only 605 pesos but the cost to the dealer in Chile was 1,693 pesos — a nominal level of protection of 180 per cent.[6]

Increasing tariffs were not the only form of protection for industry. Exchange controls were soon in force and by 1932 the Central Bank was specifically giving preference for the import of raw materials for national industry. Multiple exchange rates also came into force in 1932. Multiple exchange rates gave a great stimulus to Chilean industry. For example, in 1950, the official exchange rate was 19.37 pesos to the dollar. Imports of machinery for state and copper mining companies were often made at this rate. At this so-called 'official' rate an industrial product worth 2,000 dollars would cost only 38,740 pesos to import. But almost all other imports used the so-called 'special commercial' rate. This was equivalent to 50 pesos to the dollar. At this rate, the industrial product worth 2,000 dollars would cost 100,000 pesos. In this way, if an importer or industrialist could not import a product at the official rate, there was a great incentive for him to manufacture the product locally. Finally, the Chilean government imposed quotas on the import of manufactured goods from other countries. This was part of Chile's bilateral trading policy. After the collapse of Chile's copper and nitrate exports between 1929 and 1932, Chile attempted to build up its trade again in a series of bilateral trade agreements with other countries. This avoided any major trade imbalances and encouraged other countries to increase their import of Chilean products.

This combination of tariffs, exchange controls, multiple exchange rates and quotas produced a heavily protected national market. It proved very attractive for a wide variety of industrial entrepreneurs and industrial activity rose as a result. According to the statistical analysis of Ballesteros and Davis,[7] industrial production doubled between 1929 and 1944 and trebled between 1929 and 1952. Industrial employment more than doubled between 1928 and 1957, from 88,464 to 216,605 (see Table 11.2).

However, industrial growth varied greatly between sectors. The most impressive growth occurred in those sectors where imports had formerly dominated national consumption. For example, in 1929, imported textiles accounted for 70 per cent of the national market and imported metal products for 80 per cent of national consumption. By 1935, the imported proportions of national markets in textiles and metal products had been reduced to 25 and 40 per cent respectively.[8] This radical change in the sourcing of manufactured textiles and metal products — from North American and European companies to local manufacturing firms — formed the basis of twenty-five years of growth in production and employment. In the textile sector, production and employment increased by 900 and 600 per cent respectively between 1928 and 1950. In the metal products sector, employment increased four times over a similar period (see Figure 11.1). Similarly, rapid increases in domestic production and employment were recorded in other sectors where imports were dominant before 1929 — paper, construction materials, glass and chemicals. Slower growth was recorded in those sectors where the free-trade model had permitted national industry to supply most of local demand. In the food, drink, clothing, footwear, wood, furniture and leather sectors, national production had supplied more than 75 per cent of national demand during the 1920s.[9] Nevertheless, between 1928 and 1957, both the food and drink sectors tripled their production.

Such industrial expansion over a period of thirty years was aided by two further factors. First, the Second World War, by interrupting international trade in manufactures at the same time as boosting world demand for copper (now Chile's major export), gave a significant boost to national manufacturing. According to the figures of Ballesteros and Davis, industrial production nearly doubled between 1939 and 1945.[10] Industrial production increased most dramatically towards the end of the war when international trade in manufactures was most severely disrupted (after the entry of the USA into the war) but when demand for copper was at its highest.

Industrial expansion was further assisted by a growing involvement of government in the economy. The most significant event in this growing involvement was the creation of the Chilean

Figure 11.1: Chile: Employment Change by Industrial Sector, 1928–57

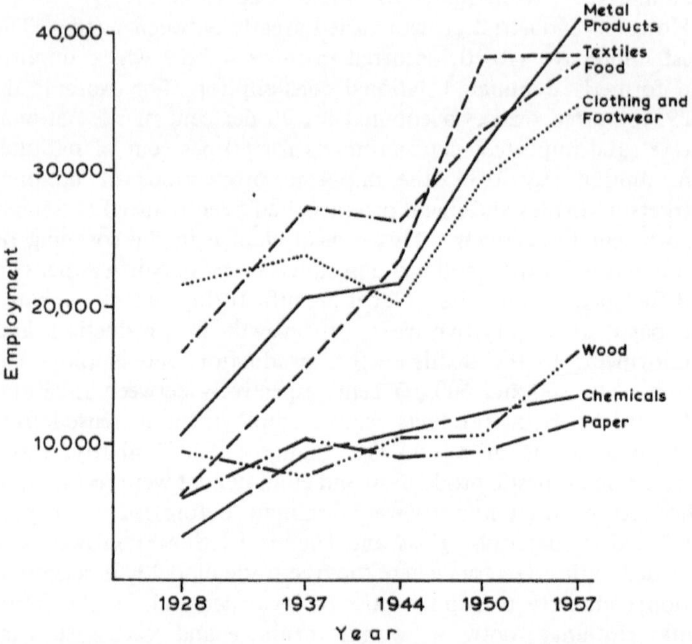

Source: Chilean Industrial Censuses for 1928, 1937, 1944, 1950 and 1957

Development Corporation (Corporación de Fomento de la Producción, or CORFO) in 1939. Financing was drawn from an earmarked portion of the 1939 income tax increase on copper mining and represented a conscious government attempt to tax the foreign-controlled export sector in order to increase the domestic formation of capital. In this way, CORFO became the key state mechanism for the planning and promotion of economic development. But while CORFO has been essentially a government-controlled enterprise, 'it has also, by including representatives from the major private economic associations on its board of governors, attempted to evolve a type of collaborative guidance for the economic system'.[11]

CORFO was granted far-reaching powers to assist in the development and execution of plans and projects. These plans could be sectoral or regional in nature. CORFO became a supervisory agency for all sectors of industry and its authorisation was required both for starting new production facilities and for expanding existing plants. To facilitate industrial development, CORFO provided technical assistance by acting as engineering consultant and market and purchasing agent. It assisted in the import of machinery and equipment, normally achieving imports at the favourable 'official' rate, and sought foreign financing for projects it approved. In its capacity as a government development bank, CORFO made loans to private firms, mixed enterprises and state corporations, and also made equity investments in partnership with private (and primarily Chilean) capital, both as a majority and as a minority stockholder. In investment areas where private interests had not ventured, CORFO exercised its power to organise and initiate companies, subsequently operating them as either state undertakings or selling them in whole or in part to private investors.

In the first twenty years of its existence, CORFO loans were granted to hundreds of enterprises in a wide variety of fields and investments were made in the stock of some 66 companies, in 41 of which CORFO held a majority interest.[12] These investments were highly diversified but certain key sectors were also identified. In the mid-1940s, CORFO began to organise a programme of hydroelectric expansion through the National Electricity Enterprise (ENDESA). This became the major investment programme of the period closely followed by investment in the National Petroleum Enterprise (ENAP) for oil exploration and production in Chile's extreme south. Among the other larger investment projects were a copper rolling mill and wire plant, the beet-sugar refineries operated by the National Sugar Industry (IANSA, established in 1952), a national hotel corporation (Hotelera Nacional, organised in 1951), an agricultural machinery and equipment company, a rubber and tyre plant, and the national steel mill. When CORFO was established in 1939, one of its main problems was to deal effectively with the lack of iron and steel production in Chile. At first, it attempted to support the faltering iron blast furnace at Corral, Valdivia, and within five years invested 48 million pesos, so that by 1945 the government effectively owned 90 per cent of the stock. However, the whole plant was too small and expensive, based on an outmoded smelting operation and with no real potential for the future. Thus, in 1943, CORFO drew up plans for a new plant at Huachipato which was to be organised by the Compañía de

Acero del Pacífico (CAP) and which would receive half of its capital from CORFO.

The decision of CORFO to build a steel mill in Chile was also due to the belief that Chile should have considerable comparative advantages in steel production. Most of the raw materials were present in Chilean territory — iron ore from mines to the north of La Serena (first El Tofo and then El Romeral), coal from Lota, limestone from the island of Guarello in southern Chile. In terms of distance, Huachipato, near Concepción, was midway between the iron ore and limestone deposits and adjacent to the local coal deposits (see Figure 11.2); in fact, local coal had to be supplemented with better quality coking coal from North America. The decision to locate at Huachipato was also influenced by the need for access to abundant supplies of electricity and water (each ton of steel required 15 tons of water). The nearby hydroelectric plant of ENDESA at Abanico provided the electricity and the river Bío-Bío the water (see Figure 11.2). The Huachipato steel mill was completed in 1950 and by 1957 had a capacity of 760,000 tons of steel and steel products. The plant supplied about 80 per cent of the Chilean market for steel and steel products in the 1950s and exported 20 per cent of its production.[13]

The three decades between 1929 and 1957 witnessed a considerable expansion of industrial activity. The combination of a highly protected market and strong government involvement in the industrial process caused a trebling of industrial production in the period. What were the spatial ramifications of the process?

One would assume that the growing involvement of government in almost all aspects of the industrial process would make the seat of government an attractive location for the industrial firm. CORFO itself was located in Santiago and the industrialist in Santiago would therefore be at an advantage over his provincial rival in terms of technical assistance, exchange rate lobbying, contacts for foreign financing, access to CORFO loans and information about new ventures and programmes that emanated from the ever-expanding CORFO bureaucracy in Santiago. Although CORFO established a number of offices in provincial towns, the organisation of CORFO was highly centralised. All new programmes and their precise terms were developed in CORFO's head office in close collaboration with the Ministry of the Economy. With governments controlling such vital aspects of the industrial process as the extent of foreign competition, tariffs, exchange

Table 11.2: Chile: The Changing Distribution of Manufacturing Employment, 1928–57

	1928		1957	
Region	Manufacturing Workers	% of total	Manufacturing Workers	% of total
Santiago	37,970	42.9	123,594	57.1
Valparaíso	16,639	18.8	19,578	9.0
Concepción	7,377	8.3	22,054	10.2
Rest of Chile	26,478	30.0	51,379	23.7
Total	88,464	100.0	216,605	100.0

Source: Chilean Industrial Censuses of 1928 and 1957

rates, subsidised loans, technical assistance, links with foreign finance, and new technical information, industrial entrepreneurs in Santiago would find it easier than their provincial competitors in two crucial areas. First, they would find it easier to develop a system of contacts in CORFO that could inform them of the policy trends in all those industrial variables controlled by government. Secondly, they would find it easier to use contacts in CORFO (alongside contacts in the two houses of Congress and SOFOFA, the industrialists' representative body) to lobby for particular proposals and changes in the state framework that they thought necessary.

Increasing industrial concentration could therefore have been assumed from the growing intervention of government in industry between 1928 and 1957. Table 11.2 demonstrates that this is precisely what happened. Santiago's proportion of Chilean manufacturing employment increased from 42.9 per cent in 1928 to 57.1 per cent in 1957. Of the twenty sectors listed in the Chilean Industrial Census of 1957, eleven had a higher level of employment concentration in Santiago than the average and fifteen had a higher spatial concentration of value-added than the average (see Table 11.3). Three of the industrial sectors that had grown most rapidly under government protection (electrical goods, rubber and metal products) had over 90 per cent of their value-added generated in the primate city. Market-oriented goods, of both traditional and modern varieties, were also significantly concentrated in Santiago. Only strongly resource-oriented sectors maintained a decentralised distribution — tobacco, petroleum and coal derivatives, food, basic metals (mainly copper in North Chile and steel in

Table 11.3: The Centralisation of Chilean Industry in the Santiago Metropolitan Area by Sector, 1957 (percentages refer to the proportion of Chilean employment and value-added in each sector generated in Santiago)

	Employment	Value-added
Electrical Goods	91.7	96.9
Rubber	90.9	95.0
Metal Products	88.2	91.7
Miscellaneous	84.0	87.9
Paper	77.5	57.3
Footwear and Clothing	76.5	79.7
Machinery	76.2	80.0
Printing	69.1	68.6
Furniture	67.0	77.4
Textiles	66.4	62.0
Leather	66.0	70.4
Beverages	52.0	54.5
Lumber	47.1	51.9
Transport Products	46.5	54.5
Chemicals	45.1	58.6
Non-Metallic Mineral Products	40.7	34.2
Basic Metals	34.7	18.5
Food	30.2	27.9
Petroleum and Coal Derivatives	8.4	1.7
Tobacco	2.5	0.2
All Industrial Sectors	57.1	48.0

Source: Chilean Industrial Census of 1957

Concepción), and construction materials.

Table 11.2 demonstrates that the retention of such resource-oriented sectors in the Chilean periphery did little to increase employment levels there. The proportion of manufacturing employment working outside the three major metropolitan areas declined from 30 per cent in 1928 to 23.7 per cent in 1957. The most notable relative decline occurred in the southern industrial town of Valdivia. As the consumer industries of Santiago expanded and transport costs between Santiago and the Chilean south continued to decrease, the capital's industry took a larger share of the southern market. Valdivia's brewing, leather and footwear industries were notable victims of the process. In 1950, the opening of Concepción's steel mill spelt the end for Valdivia's iron blast furnace and its metallurgical industry declined as a result. Between

Figure 11.2: Chile: The Manufacturing Region of Concepción in 1957

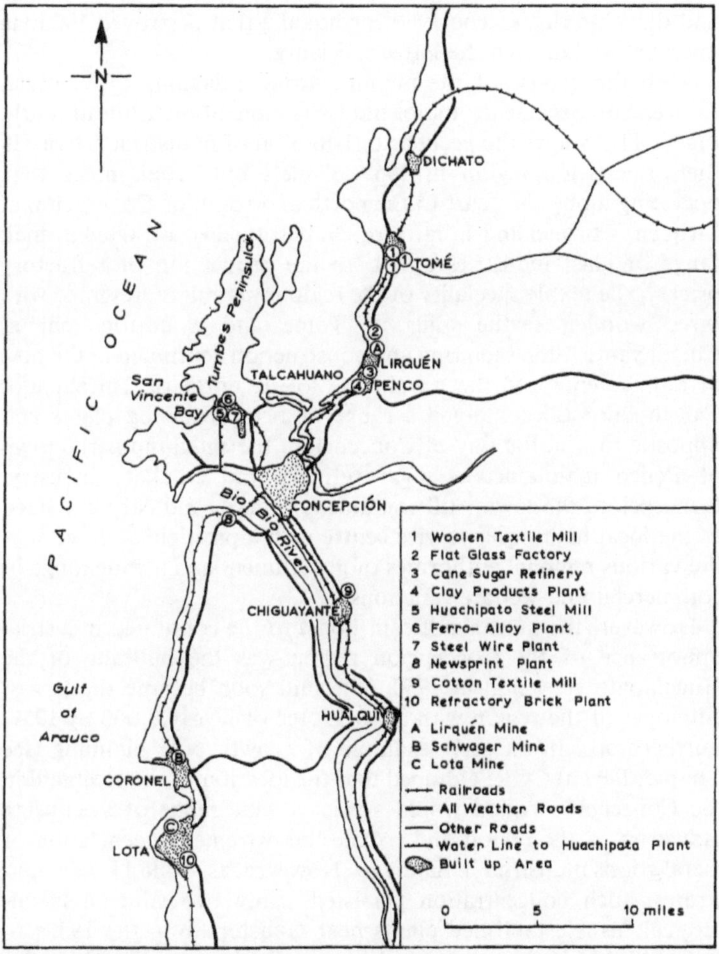

Source: J. H. Butler, *Manufacturing in the Concepción region of Chile* (National Academy of Sciences, Washington D.C., 1960)

1928 and 1957, manufacturing employment in Valdivia barely increased as four of its six industrial sectors declined. The industry that remained was intimately linked with local resources — timber and dairy products. The 1960 earthquake that destroyed Valdivia operated to end even the latter speciality.

With the demise of the manufacturing functions of Valdivia, Concepción became the major industrial area of the Chilean south. Figure 11.2 shows the peculiar distribution of industrial activity in the Concepción region in 1957. Chile's only coal mines were operating along the coast to the north and south of Concepción at Lirquén, Coronel and Lota. Lirquén had further attracted a small range of glass industries and Lota the production of refractory bricks. The textile speciality of the region was still represented with three woollen textile mills at Tomé and a cotton mill at Chiguayante. Ship repairing and construction continued at the port of San Vicente and the naval base to the north of Talcahuano. Talcahuano still contained a range of fish-processing plants and opposite to it in the Bay of Concepción, the small industrial town of Penco maintained a sugar refinery and crockery industry. Concepción, the town, still contained no major industry but acted as the local high-order service centre with a prestigious university, the various regional authorities of government and a wide range of commercial and welfare functions.

However, the most significant boost to the continued industrial importance of the Concepción region was the building of the Huachipato steel mill in 1950. The mill soon became the largest employer of the region with a workforce of nearly 6,000 by 1957. Furthermore, in an early example of growth pole planning (see Chapter Seven) CORFO hoped that the location of the steel mill in the Concepción region would attract a wide range of steel-using industries to the region and reduce the extreme concentration of metal goods industries in Santiago. However, as Table 11.3 demonstrates, such concentration persisted. Only two Chilean metallurgical firms established plants near Huachipato in the 1950s — INCHALAM built a plant producing steel wire, and CARBURO a plant producing ferromanganese. No foreign firm created a steel-using plant in the Concepción region and most Chilean metallurgical firms remained at their market location in Santiago. The strength of Santiago in the national metallurgical market can be gauged from the destination of INCHALAM's steel wire products — nearly 70 per cent were sold in the Santiago metropolitan area.[14]

Despite the lack of success of the metallurgical growth pole, industrial production and employment in Concepción grew faster than at any other location outside Santiago between 1928 and 1957 (see Table 11.2). As a result, the Concepción metropolitan region replaced the Valparaíso metropolitan region as the second-ranking industrial region of Chile. The protectionist framework of Chile's industrial development in the 1930s and 1940s did not promote industrial growth in Valparaíso (see Table 11.2). The port-industrial sectors of Valparaíso (chemicals, oil and metal refining, food processing) maintained their importance, but other industrial sectors tended to stagnate; growth was concentrated in Santiago. Industrial expansion was also constrained by a lack of flat land in Valparaíso, sandwiched as it is along a narrow coastal plain between Chile's coastal range and the Pacific. Furthermore, the adjacent northern town of Viña del Mar became the major tourist resort of Chile at this time and shunned industrial growth. New industrial plants wishing to locate in the Valparaíso metropolitan region had either to locate inland from Viña del Mar or in Quilpué. at a considerable distance from the port of Valparaíso.

Industrial Stagnation and Radical Alternatives (1957–73)

According to Ballesteros and Davis, industrial production in Chile trebled between 1929 and 1952; between 1952 and 1957, however, it virtually stagnated, increasing by only 11 per cent for the period.[15] According to Mamalakis, industrial production nearly doubled between 1940 and 1951; in the next ten years, however, industrial production increased by less than 20 per cent.[16] During the 1950s and early 1960s, a definite period of industrial stagnation can be identified. The worst period was between 1958 and 1961; industrial production in these four years was consistently below that of 1957.[17]

The most fundamental problem causing industrial stagnation concerned the external relations of the economy. For in striving for a protected economy, the emphasis had been taken away from exports. Indeed, because of the multiple exchange rate system, exports were effectively made more expensive. This was manageable for Chile's main export, copper, because Chile is the world's lowest-cost producer of this commodity. But, in other areas, such favourable cost differentials did not exist, and exports stagnated.

Even copper exports stagnated. It was not until 1959 that the country's copper production began to surpass the levels reached in the first half of the 1940s. As a result, Chile's external position became weak on two counts. First, exports were increasing very slowly. Secondly, copper's share of those exports kept on increasing, so that Chile was becoming overdependent on the export of just one commodity — and a commodity, furthermore, with a notoriously unreliable and erratic price in world markets.

Static world volume, alongside erratic prices, was a problem because the process of industrialisation in Chile had set up significant demands for the constant import of intermediate goods and raw materials for the consumer goods industries.

> Both domestic industrial production and employment levels came to depend upon the ability of the external sector to supply the new factories with a steady flow of fuel, raw materials, and intermediate goods for processing. Whereas formerly a contraction in the capacity to import had brought discomfort mainly to the upper- and middle-class consumers of imported finished goods, the consequence of foreign-exchange shortages now became even more far-reaching. In addition to the limitations imposed on the industrialization process by domestic market size, it was becoming clear that there were also limitations which might emerge from the evolving structure of imports.[18]

Three further factors have been attributed to the onset of industrial stagnation. First, there was the growing problem of inflation. Between 1954 and 1956, the average annual rate of inflation in Chile rose to 68 per cent and an annual maximum of 89 per cent was recorded.[19] Excessive increases in the money supply have been seen as a principal cause of inflation. The need for such increases was intimately linked with the process of industrialisation.

> Since the whole development undertaking — with its array of price supports, subsidies, new lines of industrial development, and welfare and housing programs — almost continuously exceeded the capacity of the system of public finance to mobilize resources except through inflationary forced savings, the result was an almost unbroken series of budgetary deficits which had to be financed out of currency expansion.[20]

The problems that high inflation brought for private-sector investment in industry together with the strong involvement of CORFO in industrial planning and development, meant that in the early 1950s 'the public sector accounted for 72 per cent of the overall low rate of gross domestic investment, a proportion between public and private shares which was almost the exact reverse of that prevailing in Latin America as a whole.'[21]

A second factor put forward to explain industrial stagnation concerned the skewed distribution of income in Chile. It was argued that such an income distribution caused industrial expansion to slow because domestic manufacturing production, although supplying a majority of the Chilean population, did not reach a substantial proportion of Chileans in the poorer section of the community.[22] The upper and middle classes, unionised industrial workers and miners, were the principal consumers of national manufacturing production. But the increasing number of urban poor working in the informal sector of the economy and large sections of the rural population of Chile were normally too poor to make significant purchases of relatively high-cost Chilean manufactures. It was argued that a redistribution of income from the well-off to the poor was needed, so the demand for manufactures would increase. Such redistribution would create the catalyst for the creation of a favourable spiral in which industrialists would respond by increasing production and thus benefit from increasing economies of scale and lower production costs. Theoretically, the industrialist would then be persuaded to reduce his prices and, in so doing, would fuel further increases in demand.

A third reason for industrial stagnation concentrated on the process of import substitution itself. Chapter Two has already described the planned sequence of import substitution from non-durable to durable to intermediate and finally to capital goods. The occurrence of industrial stagnation in Chile in the 1950s within a framework of import substitution had a profound effect on the ECLA exponents of import substitution, themselves operating from ECLA (later CEPAL) headquarters in southeast Santiago. ECLA economists came to make a distinction between easy and difficult stages of import substitution. The easy stage involved the manufacture of most non-durable goods, some durable consumer goods of simple technology and those intermediate goods linked to the processing of basic national resources — such as refined copper and steel in Chile. However, the process of import substitution could then stagnate as the difficult stage was reached. The difficult stage of import substitution involved the production of high technology goods — in the durable good, intermediate and capital good sectors. In these sectors, national entrepreneurs had little knowledge and expertise and found it difficult to compete in the Chilean market despite high levels of protection. Stronger government involvement was recommended to assist in the shift from the easy to the difficult stage of import substitution and incentives for the participation of foreign companies in the high technology sectors.

Radical solutions were required, then, for the problem of industrial stagnation and between 1964 and 1973 two governments with radical industrial strategies came to power. Between 1961 and 1971, industrial growth again took place with industrial GDP in 1971 being 78 per cent greater than that of 1961.[23]

The first government to tackle the problem of industrial stagnation was that of Alessandri (1958–64). Between 1959 and 1961, the Alessandri government had managed to bring down the level of inflation from 38.6 per cent to 7.7 per cent, but industrial production had also declined. In the second half of his administration, Alessandri began to focus on the problem of industrial stagnation. His government's interpretation of industrial stagnation closely followed that of the Economic Commission for Latin America (ECLA). As a result, the Alessandri government's main thrust was in extending the range of manufactures in Chile. Technologically advanced products, rather than being imported, were first assembled and then manufactured in Chile. Motor cars, televisions, gramophones, refrigerators and radios, to name but a few, began to be manufactured in Chile during Alessandri's administration.[24]

Stress was placed on manufacture rather than assembly as can be demonstrated by the motor vehicle legislation that the Alessandri government passed in 1962. The third section of Decree-law 835, the legislation that controlled the evolution of the motor vehicle industry through the Automobile Commission, described the process of national integration that the industry was to follow.[25] Integration in this sense referred to the progressive use by the motor vehicle company of nationally produced components. The rate at which motor car producers needed to integrate nationally produced components within their productive structure was outlined for the following seven years as Table 11.4 reveals. By 1968, virtually 53 per cent of the total FOB value of the assembled

Table 11.4: The Progressive Integration Levels Imposed on Motor Vehicle Companies Producing in Chile 1963–9

Year	Real Integration (% of FOB Price)	Assembly and Finishing[a] (% of FOB Price)	National Integration Levels
1963	20.0	10	30.0
1964	26.6	10	36.6
1965	33.4	8	41.4
1966	45.0	5	50.0
1967	50.0	5	55.0
1968	52.94	5	57.94
1969	52.94	5	57.94

a. Assembly and finishing costs refer to the cost of labour in the final assembly process, other assembly costs, the cost of soldering the bodywork, painting and general finishing
Source: Decree-law 835, 1962

car had to correspond to parts produced in Chile.

As most technologically advanced industries were spatially concentrating in Santiago, considerable concern was expressed about allowing such critical industries as that of motor vehicles to locate there. As a result, Decree-law 835 contained a special clause not permitting the manufacture of motor vehicles in the province of Santiago. Meanwhile, a wide range of tax reductions and duty exemptions on imported parts was enjoyed by the northern port of Arica.[26] As a result, the large number of firms that started to manufacture foreign motor cars in Chile (there were twenty firms already operating in 1962) located their plants in a notably decentralised location. Subsequently in 1967, lesser but still substantial advantages were given to firms locating in the three provinces adjacent to Santiago. As a result, Peugeot/Renault moved to Los Andes, Ford to Casablanca and Fiat to Rancagua (see Figure 7.1).

The Alessandri government was followed by the more radical government of Frei (1964–70). The last three years of Alessandri's government had witnessed a 23 per cent increase in industrial production — more than the previous decade — and Frei was intent on maintaining such industrial growth. He therefore continued the policy of widening Chile's product range. At the same time, the Frei government aimed to expand industrial production through a gradual process of income redistribution. The combination of a wage policy favouring the poorer sections of the community and an anti-inflation policy were basic to this. According to Muñoz, these

Table 11.5: Chile: Increases in the Proportion of Foreign Investment in Certain Manufacturing Sectors, 1967–9

	1967	1969
Tobacco	59.1	58.6
Electrical Products	48.9	59.9
Rubber	44.2	45.1
Chemicals	31.1	38.3
Construction Materials	25.6	24.7
Footwear and Clothing	23.1	18.7
Basic Metals	16.7	13.6
Metal Products	16.5	18.4
Drinks	15.4	24.6
Transport Products	10.7	43.8
Machinery	7.1	14.7
Paper	2.7	16.4
Printing and Publishing	1.7	15.2
Total Manufacturing Industry[a]	16.6	20.3

a. Includes all industrial sectors

Source: L. Pacheco, La inversión extranjera y las corporaciones internacionales en el desarrollo industrial chileno, in O. Muñoz (ed.) *Proceso a la industrialización chilena* (Universidad Católica, Santiago, 1972) 106–55

policies caused the percentage of workers with wages less than that of the minimum wage to decline from 40 per cent in 1961 to 20 per cent in 1967. Furthermore, the percentage of blue collar workers enjoying a wage more than three times the minimum rose from 8 per cent in 1961 to 21 per cent in 1967.[27] From such figures it can be assumed that a modest redistribution of income was engineered in the first three years of Frei's government.

In order to improve the technological expertise of Chilean industry in the 'difficult' stage of import substitution, the Frei government welcomed foreign investment in manufacturing and the establishment of manufacturing subsidiaries by foreign corporations. As a result the years between 1964 and 1969 became the most significant period for direct foreign investment in manufacturing that Chile has yet witnessed. It should be remembered that the Frei government was popular with the US administration of Johnson as it represented the US ideal of a democratic, reforming government in Latin America. For example the Frei government instigated wide-ranging agrarian and educational reforms. Consequently, there was considerable encouragement for US corporations to invest in Chile.

Table 11.6: Foreign investment in the Hundred Principal Private Companies of Chile in 1969

	Number of firms	Number of firms with foreign investment as a % of total investment between		
		1–50%	50.1–100%	$\frac{B+C}{A} \times 100$
	A	B	C	
Food	15	6	3	60.0
Drinks	4	1	1	50.0
Tobacco	1	–	1	100.0
Textiles	16	6	1	43.8
Leather and Footwear	5	–	2	40.0
Wood	4	–	1	25.0
Paper	3	–	1	33.3
Printing and Publishing	3	1	–	33.3
Chemicals	11	3	6	81.8
Rubber	1	–	1	100.0
Construction Materials	7	5	2	100.0
Basic Metals	6	6	–	100.0
Metal Products	5	1	3	80.0
Machinery	3	1	1	66.7
Electrical Products	7	2	2	57.1
Transport Products	5	1	3	80.0
Others	4	–	–	–
TOTAL	100	33	28	61.0

Source: L. Pacheco, La inversión extranjera y las corporaciones internacionales en el desarrollo industrial chileno, in O. Muñoz (ed.) *Proceso a la industrialización chilena* (Universidad Católica, Santiago, 1972) 106–55

Between 1967 and 1969 alone, the proportion of foreign investment in Chilean manufacturing increased from 16.6 to 20.3 per cent[28] (see Table 11.5). In the technologically more advanced sectors, however, increases in the proportion of foreign investment were more substantial. In the transport products sector (mainly motor vehicles), the proportion of foreign investment quadrupled (see Table 11.5). The participation of foreign investment increased substantially in six other sectors — electrical products, chemicals, drinks, machinery, paper and printing and publishing. Reduction in foreign participation occurred in traditional sectors (footwear and clothing) or in sectors where government was becoming involved (basic metals as a result of Frei's 'Chileanisation' of the national copper industry). Not only was foreign investment prominent in the technologically advanced sectors, but also in the largest companies. Table 11.6 demonstrates the level of foreign investment in the hundred principal private companies of Chile in 1969. Foreign companies had investments in 61 of the leading hundred companies, although they had a majority shareholding in only 28. However, as Pacheco pointed out, Chilean companies with foreign investment in a minority could be controlled by foreign interests due to divisions between Chilean shareholders.[29] Thus, by 1969, foreign investment in Chilean manufacturing was substantial. Foreign interests controlled approximately one third of Chile's largest private companies and they were particularly prominent in the faster-growing sectors.

Partly as a result of income redistribution and partly as a result of a further widening of Chile's policy of 'autonomous' manufacturing, industrial production grew by a total of 19 per cent in the first three years of Frei's administration. However, industrial expansion began to slow once again in the last three years of his administration when a total growth of only 7 per cent was recorded.[30] The origins and causes of industrial stagnation once again became a crucial area of political debate. For the winner of the 1970 Presidential election — Salvador Allende — there was little doubt where the problem lay.

The 1970 Chilean election still presents the only case of a Marxist-Leninist candidate achieving power through the ballot box. The policies that ensued were suitably radical. In terms of industrial policy, there were two principal aims.[31] First, Allende argued that it was necessary to reduce the impact of 'monopoly capitalism' on Chile's industrial activity. Monopoly capitalism in this sense referred to foreign capital and the large Chilean firms that were influential in individual industrial sectors. Secondly, industrial stagnation would be solved by a massive redistribution of income. Demand would increase, full utilisation of industrial capacity would be achieved, the unit costs of industry would come down — and with close government supervision, prices would come down as well.

The reduction in the impact of monopoly capitalism was to be achieved by the state taking over all foreign companies and local 'monopolies' that enjoyed a pre-eminent position in one or more industrial sectors. According to Allende's manifesto, the state would expropriate all those enterprises whose value was greater than 14 million escudos ($2 million) in 1969. According to this definition, 250 enterprises were due for a change of ownership, but by the end of 1971, the names of 90 enterprises were listed for nationalisation or for the state to take a majority shareholding.[32] Unfortunately for the Popular Unity government of Salvador Allende, there was no 'legal' way of expropriating and nationalising these industries. Because the Popular Unity coalition had achieved only 36 per cent of the vote at the 1970 election, it did not have a majority in Congress and could not pass a nationalisation bill. It therefore had to adopt other ways of achieving its aims of taking over industry. Three such techniques can be identified.

The principal method was intervention. This was an instrument which Eduardo Novoa, Allende's legal adviser, had found when researching into the legislation of the short-lived 'Socialist Republic' of 1932. Decree-law 520 provides for 'the intervention of the central power in all industries producing basic necessities which infringe on laws of functioning freely established by the administrative authorities'. The principal tactic used was that of creating some serious labour disturbance in a particular factory. This would inevitably lead to a decline in production, sometimes even a total stoppage, which would therefore 'infringe on laws of functioning'. A government inspector would then be appointed to make a report on the situation. If the report was detrimental, a state *interventor* would be appointed to take over the immediate running of the company with total control over all facets of the industrial firm in question. The owners or shareholders could appeal to the courts, but if the latter passed in favour of the owners, the government could resort to a 'decree of insistence', a decree signed by all members of the Cabinet which overruled the decision of the Appeal Court. This is an excellent example of the wide range of powers and prerogatives enjoyed historically by the Chilean Executive[33] and their extension under Allende. According to the Economics Institute of the University of Chile, between November 1970 and December 1971, 134 firms suffered intervention, of which only 42 were returned to their owners. In 1972, 77 firms had *interventores* appointed, and only 15 were later returned to their original owners.[34] With the March elections of 1973, the rate of intervention slowed. However, after the abortive uprising of a tank regiment on 29 June 1973, Allende called on the workers to take over all industrial establishments; 526 firms, of which at least 100 were medium-sized, were taken in the aftermath.[35] It effectively provided the point when the great majority of manufacturing industry came to be controlled by the state.

The second method of taking over industry was that of requisition, whereby the government would requisition a firm under a state of emergency, on the pretext, for example, of falling behind in its production quotas. The requisition became the main tool used against the 90 listed industries and it was in this respect that the 'decree of insistence' was most commonly used. Between November 1970 and the end of 1972, 126 industrial establishments were requisitioned by the government, of which 39 were contained in the list of 90.

Thus, the Popular Unity government was forced to requisition 39 of the 49 firms on the list of 90 which they took before the March 1973 elections. The other ten were either obtained by intervention or by negotiation. However, on 10 April 1973, the Cabinet issued a 'decree of insistence' expropriating the remaining 41 private companies included in the list of 90 on the grounds that they failed to meet production quotas.

The third method was by negotiation and purchase of shares. To this end, the Allende government turned the state planning agency, CORFO, into a giant holding company and used public money to buy a controlling interest in private companies. Industrial entrepreneurs had two main motives for negotiation. First, there were those who adhered to the view that a quick surrender was the best policy for the industrialists, given the aim of the government to nationalise the greater part of industry. By December 1972, 36 industrial firms with no prior experience of conflict had completed negotiations for the transfer of the industry to the government, while another 18 firms were in the process of negotiation. These firms included firms in the car industry (e.g. CORFO-Citroën, CORFO-Peugeot and 3 major parts industries — Femsaco, ENATAP and Cormecanica), firms producing agricultural machinery, and some metallurgical firms. Second, there were those industrial entrepreneurs who had suffered conflict due to either intervention or requisition and who thought negotiation was the best way of cutting their losses. Thus by the end of 1972, 45 firms which had previously experienced intervention or requisition had come to financial terms with CORFO or were in the process of doing so.

Industrial production responded erratically to Allende's policies. In 1971, industrial production, benefiting from the redistribution of income, rose by 14.6 per cent.[36] However, by 1972, industrial growth slowed down considerably as production increased by only

Table 11.7: The Changing Distribution of Manufacturing Employment, 1957–67

Province	1957	1967
Santiago	57.1[a]	56.9
Concepción	10.2[a]	9.9
Valparaíso	9.0[a]	8.7
Tarapacá	0.9[b]	2.6
Rest of Chile	22.8	21.9
Total	100.0	100.0

a. Figures refer to spatial units less than that of the province
b. Estimate
Source: Industrial Censuses of 1957 and 1967

2.8 per cent for the year. Allende's combination of industrial policies was beginning to show a major flaw — lack of investment. For in taking over and purchasing a wide range of industry, huge sums of public money were required merely to buy the firms and keep them running; there were no public resources left for new investment. Foreign investment had been largely expropriated and private Chilean investment was not forthcoming due to widespread requisition and intervention. Indeed, to keep in line with its commitments, the Allende government had to increase the money supply severely. Between December 1970 and December 1971, the money supply expanded by 116 per cent; between December 1971 and December 1972 by 165 per cent; and between July 1972 and July 1973, as the economy went increasingly out of control, the money supply increased by 287 per cent. Such extraordinary increases in money supply fuelled inflation which rose from 15.5 per cent in September 1971 to 323 per cent in July 1973. By 1973, industrial production was in decline. A comparison of industrial production for the months January to August for both 1972 and 1973 show an overall decline of 5.4 per cent and a comparison of the last four months of Allende's government with the corresponding four months of 1972 show a 9 per cent decline in production. On 11 September 1973 Allende's government came to a violent end as a military Junta organised a coup and took power.

Despite such dramatic changes in government and industrial policy between 1957 and 1973, little change occurred in the spatial pattern of industry (see Tables 11.7 and 11.9). Detailed analysis, however, is prevented by a poor regional coverage of industrial data for the period. One problem is that the regional definitions of

Table 11.8: The Spatial Centralisation of Selected Industrial Sectors in Santiago Province in 1967 Based on Employment and Value-added Totals (percentages)

	Employment	Value-added
Pharmaceuticals	98.3	99.2
Electrical Durable Goods	96.4	97.1
Plastics	95.3	96.9
Rubber	95.1	98.4
Professional and Scientific Equipment	90.3	85.3
Metal Products	83.5	87.2
Footwear	82.8	85.5
Clothing	79.2	84.0
Miscellaneous	77.8	77.5
Glass	77.8	74.5
Textiles	76.6	77.1
Leather	75.6	85.3
Furniture	69.3	78.3
Printing and Publishing	68.4	79.1
Motor Vehicles	68.4	36.3
Electronic Products	67.1	60.4
Chemicals (resins, fibres, paints etc.)	66.5	68.8
Machinery	66.2	78.2
Constructional Products	61.1	68.9
Paper	55.6	43.8
Electrical Machinery	43.8	54.8
Ceramics	38.0	36.0
Copper and Other Non-Ferrous Metals	37.7	6.1
Food	36.8	30.1
Beverages	36.0	49.0
Iron and Steel	32.6	24.1
Shipbuilding and Railways	30.7	34.7
Lumber	18.9	25.3
Petroleum and Coal Derivatives	12.7	11.9
Tobacco	1.3	0.2

Source: 1967 Industrial Census

the 1967 Census are different from those of the 1957 Census. Meanwhile, regional coverage of manufacturing from 1970 to 1973 was restricted to firms with over fifty employees.

Between 1957 and 1967, it is evident that the strong spatial centralisation of industry was maintained. Despite government schemes of industrial decentralisation, Santiago still contained 57 per cent of Chilean manufacturing employment in 1967 (see Table 11.7). Moreover, the 1967 Industrial Census shows that the new growth industries of Chile in the 1960s, the technologically

Table 11.9: The Changing Distribution of Manufacturing Employment in Medium and Large Plants (Over 50 Employees) Between 1967 and 1973 (percentages)

Region	1967	1970	1971	1972	1973
Santiago	59.8	60.8	57.3	56.3	56.9
Concepción	15.1	15.2	17.7	16.7	15.3
Valparaíso	10.4	9.7	9.6	10.1	9.9
Tarapacá	3.0	3.2	3.5	4.1	4.0
Other Regions	11.7	11.1	11.9	12.8	13.9
Total	100.0	100.0	100.0	100.0	100.0
Total Employment	233,543	244,265	247,612	259,710	264,972

Source: Industrial Census of 1967, Industrial Surveys of 1970, 1971, 1972, 1973

advanced sectors, are highly concentrated in the primate city (see Table 11.8). In particular, the pharmaceuticals, electrical goods, plastics, rubber (tyres) and professional and scientific equipment industries had over 90 per cent of their employment generated in Santiago. Even those sectors that government had attempted to decentralise, motor vehicles and electronics, had over 60 per cent of their employment in the primate city. In terms of the motor vehicle industry, this was because the motor vehicle components industry that grew up as a result of the policy of national integration was strongly concentrated in Santiago.[37] This accounts for the relatively low amount of value-added generated in Santiago in this sector in 1967 as the assembly industry, characterised by high value-added, was still concentrated in Arica. As regards the electronics industry, its high concentration in Santiago demonstrates that the governments of Alessandri and Frei only managed to decentralise the television industry;[38] most other subsectors of the electronics industry remained heavily concentrated in Santiago.

Technologically advanced industries tend to operate through medium and large firms rather than through small firms. If one examines the spatial pattern of employment for the medium and large firms with over fifty employees, it is evident that in 1967 spatial concentration of employment in these firms was higher than for that of all firms (see Table 11.9). Nearly 60 per cent of employment in firms of over fifty employees was concentrated in Santiago in 1967. In the last three years of Frei (1967–70), there was a further concentration of employment in larger firms in Santiago. A small but increasing degree of decentralisation occurred in the three years of Allende (see Table 11.9).

It has been demonstrated how Chilean industry became increasingly protected between 1957 and 1973. A major regional beneficiary of such protection was Concepción with its combination of metallurgical, textile, glass, food, ceramics, paper and timber industries. It maintained its relative position in Chile between 1957 and 1967, but achieved a greater proportion of regional growth between 1967 and 1973. Unfortunately the published statistics for this period do not divorce Concepción from the wider region and thus include the provinces of Ñuble, Arauco, Bío-Bío and Malleco. Meanwhile, the third industrial area of Chile, Valparaíso, seemed to continue its slow relative decline between 1957 and 1973.

The major case of industrial decentralisation between 1957 and 1967 concerned the government policy of decentralising technologically advanced industry to the northern port of Arica in the province of Tarapacá. In 1952, President Ibañez had made Arica a free port in order to stimulate the commerce of a town that was seeking closer links with Bolivia. At that time Arica had a population of 15,000. In 1958, President Ibañez created a regional development body, the *Junta de Adelanto de Arica*, before he left office. In the Alessandri administration, legislation was passed to concentrate the Chilean motor vehicle industry there. Under the Frei administration, legislation was passed to concentrate the electronics industry there. Firms producing televisions rapidly decentralised to Arica so that by 1972, six assembly and five components firms were producing 166,000 televisions a year and employing 3,000 workers.[39] By 1972, Arican industry employed 9,000 workers and the town had become the fourth industrial town of Chile; its population had risen to 120,000. However, apart from the case of Arica, rapid manufacturing growth occurred only in those small towns near Santiago. In the periphery as a whole, manufacturing growth was limited.

A Return to a Free-trade Model (1973–85)

In the debates on the reasons behind industrial stagnation in the 1950s, there had been a school of thought that pointed to excessive protection and government intervention as the root cause. With excessive protection, it was argued, entrepreneurs were not interested in cutting costs. Therefore, they produced high-cost products for a small domestic market. There was little possibility of exports due to the high cost nature of the products. Industrial stagnation set in, it was argued, because the efficient producers could not expand out of supplying the domestic market while the inefficient producers were cosseted from external competition by high tariffs. The solution, therefore, lay not in redistribution of income or a widening of the national product range, but in a reduction in industrial protection and associated government intervention.

This school of thought has dominated economic and industrial policy in Chile ever since the 1973 coup that placed a military regime under General Pinochet in power. Some have argued that the 1973 coup provided such a radical turnaround in economic policy that the Pinochet rather than the Allende government constituted a revolutionary change:

> For, what is currently in train in Chile is nothing less than a revolution against not only the avowedly interventionist role of the State pursued by the Allende government in national economic management, but also the long-term trends in that direction for several decades before that particular regime came to power through the ballot box. In this sense, the Allende government was more evolutionary than revolutionary in Chile's historical experience: it is its successor which is pursuing the revolutionary path.[40]

Most of the economists who have been in charge of the Chilean economy since 1973 have been civilians, many were educated at the Economics Department of the Catholic University in Santiago and some received further academic training at the University of Chicago, where Professor Milton Friedman's free-market and monetarist views prevail — hence the term 'Chicago boys' to describe the civilian economists of General Pinochet.

However, the central economic policy of less protection and freer trade dates back to the first Treasury Minister who was a member of the armed forces. On 7 January 1974 Vice-Admiral Gotuzzo outlined the economic strategy for the following three years, and a reduction in industrial protection and the promotion of more diversified exports were the cornerstones of his policy.[41] Subsequently a timetable was fixed for the progressive lowering of tariff barriers on imported manufactures to a general level of 10 per cent by 1979. Given that in 1973, it was calculated that some industries benefited from an effective protection of 10,000 per cent, the revolutionary impact of such proposals can be appreciated. Furthermore, the various economic teams of the Pinochet government reduced the role of CORFO in industrial and economic development. Virtually all the enterprises taken over by the Allende government were sold back to the private sector and some of the enterprises created by CORFO previous to 1970 were also sold off. Concomitantly, the Pinochet government embarked upon a deliberate policy of export expansion, particularly in those products where Chile's natural advantages could be exploited internationally — forestry products (timber, pulp, paper, wood chemicals), fishery products (fishmeal, fish oil, canned and frozen fish), wine, fruit and vegetables.

The industrial history of Chile subsequent to September 1973 can be conveniently divided into four parts. The first period (1973–1976) was characterised by extreme austerity and as a result industrial production declined by 27 per cent for the period.[42] A whole series of austerity measures was needed in order to reduce the 500 per cent levels of inflation of 1973 and to limit overseas borrowing. Despite levels of protection being lowered, the Chilean market was still largely supplied by manufactured products produced in Chile. Industrial decline was basically due to a dramatic reduction in internal demand, consequent upon government policies.

A period of austerity was followed by a period of tremendous growth in the Chilean economy. Between 1976 and 1979, the GDP index rose by an annual average of 7.5 per cent and reached 6.5 per cent in 1980. At the time, such dramatic growth was referred to as the 'Chilean miracle'. However, manufacturing activity was undergoing radical change, as almost all tariffs were reduced to 10 per cent by 1979 and as a process of continual devaluation kept the Chilean peso low against international currencies and boosted the export sector of manufacturing. Table 11.10 serves to summarise the dramatic changes that took place in Chilean manufacturing between 1973 and 1980. After the 27 per cent reduction in industrial activity between 1973 and 1976, industrial production had crept back to 1973 levels by 1980. However, the import-substituting sectors that had previously been the cornerstone of Chilean manufacturing suffered major production losses. The textile sector that had been the second largest manufacturing sector in Chile in 1973

(10.7 per cent of total production) was reduced to 38 per cent of its former size by 1980 — and thus became the eighth-ranking industrial sector. Regional textile specialities terminated as a result — most notably in the case of Tomé where the three woollen textile mills found that a 10 per cent protection level against foreign competition was not sufficient. The ceramics sector fared even worse than the textile sector, production falling by 74 per cent for the period, and the regional speciality of Penco, near Concepción, terminated as a result. The 1980 production of four other import substituting sectors also declined to less than half of their former levels — leather, footwear, transport products and miscellaneous.

The motor vehicle industry accounted for 75 per cent of the transport products sector in 1980 and had been made a special case in terms of tariff reductions. In order to widen the number of people able to buy a motor car, a ten per cent tariff was charged on the import of all vehicles under 850 c.c. The import of vehicles with larger engine size, however, incurred much higher tariff payments. As a result, the import of small cars boomed. By 1980, 33,387 cars with less than 850 c.c. were being imported, alongside the import of only 17,903 cars with greater engine size. This left the national industry to enjoy its peak production year of 25,208 cars. However, the percentage of national parts integrated into production had dropped from 58 to 30 per cent, so the 1980 components industry was much smaller than that of 1973.

Eight other industries that had grown rapidly in the highly protected phase of Chile's industrial growth recorded a production decline of between 50 and 75 per cent in the period 1973–1980 — chemicals, rubber, plastics, glass, metal products, machinery, electrical and electronic products, professional and scientific equipment. As regards the electronic sector, some industries, such as those producing televisions and radios, virtually disappeared altogether. Other sectors also contained industries where major decline occurred.

It is an interesting statistical point that in Table 11.10, no industrial sector (at the three digit level) declined moderately (i.e. between 1 and 25 per cent) over the period 1973 to 1980. Industrial sectors either declined more severely or increased their production. It would be tempting to see this statistical gap as the different performance of import-substituting industry on the one hand and export-oriented industry on the other. However, closer examination of the expanding industry makes such a distinction less clear.

Table 11.10: Changes in Industrial Production and Employment in Chile by Sector, 1973–80 (1973 = 100)

Sector	1980 Production (1973 = 100)	1980 Employment (1973 = 100)	Change in Production per Employee, 1973–1980 (1.00 = No change)
Food	103.9	106.7	0.97
Drinks	159.6	91.6	1.74
Tobacco	109.1	72.2	1.51
Textiles	38.0	57.7	0.66
Clothing	102.6	124.4	0.83
Leather	31.5	62.8	0.50
Footwear	47.8	66.4	0.72
Timber	109.0	109.8	0.99
Furniture	166.7	99.2	1.68
Paper	139.8	86.1	1.62
Printing and Publishing	112.2	95.4	1.18
Chemicals	74.0	74.0	1.00
Products from Coal and Oil	275.2	83.8	3.28
Rubber	51.3	69.2	0.74
Plastics	60.2	100.4	0.60
Ceramics	25.8	60.5	0.43
Glass	55.6	48.5	1.15
Construction Materials	158.4	74.3	2.13
Iron and Steel	121.1	63.0	1.92
Non-Ferrous Metals	186.2	110.8	1.68
Metal Products	72.3	86.4	0.84
Machinery	57.9	55.5	1.04
Electrical and Electronic Products	54.0	52.0	1.04
Transport Products	49.5	40.6	1.22
Professional and Scientific	55.6	83.6	0.67
Miscellaneous	47.6	42.1	1.13
TOTAL	100.5	78.3	1.28

Source: R. N. Gwynne, *Production and Spatial Change in Chilean Industry, 1974–1984* (ESRC Research Report, 1985)

Industrial sectors can be deemed to have recorded an expansion of production if their 1980 value of production was greater than their 1973 value (at 1980 prices). However, for some industries, such as the oil refining industry, increasing value of production was caused more by an increase in the value of the raw materials than an increase in the level of production. Thus the near trebling of the value of production in the coal and oil products sector is mainly due to the quadrupling in oil prices between 1973 and 1980. Furthermore, it must be remembered that in 1979 and 1980, the Pinochet government encouraged the construction sector to be the leading motor of economic recovery. Constructed square metres and the construction industry's labour force increased very rapidly in these two years. As a result, it would be expected that industrial sectors related to construction had higher production levels in 1980 than 1973, a year of below average construction. Four industrial sectors appear to have benefited from the construction boom in 1979 and 1980 — construction materials, iron and steel, furniture and timber. The timber industry also expanded to meet export demand. But only three other sectors similarly expanded to meet export demand — non-ferrous metals (mainly copper refining), paper and food (mainly related to fish — such as fish meal, fish oil and canned fish). The four other sectors that expanded production between 1973 and 1980 had negligible exports.[44] They therefore represent four sectors oriented principally to the domestic market that were able to successfully withstand increased foreign competition — printing and publishing, clothing, tobacco and drinks.

Table 11.10 refers to those industrial enterprises which have more than fifty employees. It is interesting to note that while production marginally increased among this size of firm between 1973 and 1980, employment declined by 21.7 per cent. At this point it is worth referring to Massey and Meegan's study of firms adapting to recession, and an associated reduction in demand.[45] According to Massey and Meegan, firms principally react in two ways, both of which serve to reduce employment. First, the firm can simply react by reducing capacity; as a result both production and employment decline. Alternatively, the firm can react by investing in new machinery and equipment and thereby increasing capacity. If the firm manages to achieve high production levels, the impact of economies of scale means that the unit cost of the product is reduced, prices cut and markets expanded. However, the new machinery is normally labour-saving and hence although production is increased, employment is reduced. If one translates firm behaviour into sector performance, Table 11.11 demonstrates that the performance of 21 out of Chile's 26 industrial sectors fell into one of these two categories between 1973 and 1980. Thirteen sectors suffered from 'rationalisation' and an associated decline in both production and employment; most of these sectors had expanded behind high protective walls to supply the Chilean market and found it difficult to compete against foreign competition.

Table 11.11: Chilean Industrial Sectors Classified According to Change in Levels of Production and Employment Between 1973 and 1980

		Employment		
		Increase	Decrease	
Production	Increase	Food Clothing Timber Non-Ferrous Metals	Drinks Tobacco Furniture Paper	Printing and Publishing Coal and Oil Products Construction Materials Iron and Steel
	Decrease	Plastics	Textiles Leather Footwear Chemicals Rubber Ceramics Glass	Metal Products Machinery Electrical Products Transport Products Scientific Equipment Miscellaneous

Source: R. N. Gwynne, *Production and Spatial Change in Chilean Industry, 1974–1984* (ESRC Research Report, 1985)

Meanwhile the process of investment and technical change seems to have been the rule for eight sectors, where production increased but employment decreased between 1973 and 1980 — apart from the paper industry, all these industries were primarily geared to the domestic market. The implication is that such sectors had to modernise rapidly (and in so doing reduce employment levels) in order to meet successfully the threat of foreign competition in the Chilean domestic market. It was only in the export-oriented sectors (food, timber and non-ferrous metals) and the labour-intensive clothing industry, that both production and employment increased between 1973 and 1980.

Thus, in 1973, the economic philosophy that had surrounded Chilean industry for nearly fifty years began to change radically. Between 1928 and 1973, a philosophy of industrial protection prevailed and a pattern of 'accumulating protection' was evident. In the following seven years, however, this complex and complicated framework of protection was effectively dismantled and Chile became 'perhaps the most open market in the world'.[46] But it is also evident that in making this radical switch from protection to free trade, total industrial production did not increase. As the Chilean economy itself was growing at over 7 per cent a year between 1976 and 1980, this manufacturing stagnation meant that the importance of industry to the Chilean economy nosedived dramatically; in 1974, the industrial sector accounted for 29.5 per cent of GDP but by 1980 it constituted only 21.5 per cent.[47] As the free-trading model anticipated, the greatest demise occurred in those import-substituting sectors that found it difficult to compete against foreign imports at low levels of protection. The thirteen import-substituting sectors that were identified as finding it difficult to compete against foreign imports, and hence underwent the process of rationalisation (see Table 11.11) recorded an astonishing overall decline in production of 46.4 per cent between 1973 and 1980. Overall employment decline in these thirteen setors was in fact somewhat less at 38.8 per cent for the period.

The problem of employment generation in manufacturing under a free-trade model was further demonstrated by the performance of the seven industrial sectors that grew most rapidly between 1973 and 1980 — coal and oil products, non-ferrous metals, drinks, construction materials, furniture, iron and steel and paper. While these seven sectors recorded an overall production growth of 81.7 per cent between 1973 and 1980, employment actually declined by 16 per cent. Furthermore, in 1980, the average value of production that corresponded to each worker in these seven sectors was as much as US $156,383 — or four times that of the average value of production per worker in the thirteen import-substituting sectors that were in decline. Thus, in the transformation of Chile's industrial structure between 1973 and 1980, capital-intensive sectors were replacing more labour-intensive sectors in terms of relative importance. Furthermore, the capital-intensive sectors substantially increased their capital/labour ratio in the period.

In this way, the actual results of Chile's industrial restructuring were distinctly different from those predicted by free trade theorists. As was noted in Chapter Two, international trade theory sees manufacturing growth in general and export growth in particular as emanating from labour-intensive industries in less developed countries, basically because labour costs are cheaper than those in developed countries. However, when Chilean economic ministers decided to promote manufacturing growth in Chile through an increased approximation to the world trading system, the reverse phenomenon occurred as labour-intensive industries declined and capital-intensive industries grew. There would appear to be three reasons behind this. First, many countries in East Asia were already powerful exporters of labour-intensive goods and often had cheaper labour costs than those in Chile. Thus, between 1973 and 1980, imports of cheap manufactured products from the low labour cost countries of China, South Korea, Hong Kong and Taiwan increased by a factor of thirty — from US $8 million in 1973 to US $243 million in 1980. Secondly, in terms of more sophisticated consumer goods, Chilean industrial firms did not have the technology or expertise to compete with imported manufactures from Japan. As protection levels lowered, Chilean consumers switched readily from the purchase of national products to imported (but particularly Japanese) equivalents. Japanese imports increased from US $54 million in 1973 to US $608 million in 1980 as a result[48] and Japan became second only to the United States as a source of Chilean imports. Meanwhile, increases in Chilean industrial production were mainly linked to the refining and processing of basic raw materials — copper and oil refining, steel and cement production, paper making and drinks manufacturing. Most of these industries had always had a high capital to labour ratio. In order to expand production, an increase in capital and plant rather than labour was necessary.

Up to 1980, however, some successes could be recorded for Chilean manufacturing exports. Chilean economic ministers were intent on promoting exports and on keeping the peso low against the dollar and other international currencies. As a result, manufactured exports increased by a factor of seventeen between 1973 and 1980 from US $89 million to US $1,559 million. However, from 1980 a surprising third stage in the economic management of the Pinochet government can be identified. One philosophical cornerstone of Chilean economic policy under the 'Chicago boys' had been to let the free play of the market allocate resources as far as possible. But then towards the end of 1979, it was decided to withhold the free play of the market from one critical area of the economy — the exchange rate. It was decided, in short, to hold the Chilean peso fixed against the most powerful currency in the world, the dollar, regardless of market fluctuations in world currency. As Chilean inflation was consistently higher than US inflation, the fixed exchange rate policy came to have a disastrous effect on the economy as a whole and on manufacturing in particular. For Chilean manufacturing, the fixed exchange rate policy meant that it was increasingly cheaper and easier to import than to produce. An

Table 11.12: Chile: Imports of Manufactured Consumer Goods, 1973–83 (Millions of US Dollars)

Year	Actual Imports	'Normal'[a] Imports	Difference between actual and 'normal' imports
1973	651	651	–
1974	562	774	– 212
1975	479	845	– 366
1976	443	884	– 441
1977	789	938	– 149
1978	1,078	1,011	+ 67
1979	1,327	1,138	+ 189
1980	2,070	1,154	+ 916
1981	2,786	1,259	+1,527
1982	1,471	1,287	+ 184
1983	1,022	1,325	– 303

a. The Chilean Central Bank's classification of 'normal' imports of consumer goods for the period refers to the value of consumer good imports for 1973, annually corrected in line with the US inflation rate
Source: Chilean Central Bank

imported product that was five per cent more expensive than the national product at the beginning of 1980 was five per cent cheaper by the end of that year. As a result, from late 1980 to early 1982 a massive influx of manufactured imports took place. The worst years for Chilean manufacturers were 1980 and 1981 as Table 11.12 demonstrates. This table takes 1973 manufactured imports of consumer goods as 'normal', and calculates 'normal' imports for the following years in accordance with the US inflation rate. It can be seen that in the austerity years of 1974 to 76, imports of consumer goods were considerably less than would be expected from a projection of 1973 imports. However, by 1979, real imports of manufactured consumer goods were becoming significantly greater than those regarded as normal. By 1981, the real value of consumer goods being imported was as much as $1,527 million above that regarded as normal.

Such a flood of manufactured imports in 1980 and 1981 meant that Chilean industrial production in 1981 stagnated after five years of growth. In 1982, industrial production actually fell by 16 per cent (see Table 11.13). The fixed exchange rate policy was principally to blame. First, with a high peso relative to a strong dollar, exports became expensive and manufactured exports found themselves uncompetitive in many markets. As a result, manufactured

Table 11.13: Chile: Indices of Manufacturing Growth, 1979–June 1984

1979	whole year	100
1980	,, ,,	106.5
1981	,, ,,	107.4
1982	,, ,,	90.1
1983	,, ,,	94.5
1982	January–June	89.6
1983	,,	91.7
1984	''	101.8

Source: Instituto Nacional de Estadísticas, *Indice de Producción y Venta Fisica de Industrias Manufactureras* (Santiago, July 1984)

exports declined by nearly 25 per cent between 1980 and 1982. Even worse, the boom in the import of consumer goods started to kill off large numbers of firms that had successfully managed to adapt to 10 per cent levels of protection. However, they found it impossible to compete with imported goods that were now effectively being subsidised by the fixed exchange rate policy. The thirteen import-substituting sectors that had declined by an average of 46.4 per cent between 1973 and 1980 bore the brunt of the boom in consumer good imports. All thirteen sectors recorded further production decline between 1980 and 1982 with an unweighted average decline of 36 per cent.[49]

The fixed exchange rate policy came to an end in June 1982. In the next six months a devaluation of 90 per cent occurred.[50] The return to a market rate for the Chilean peso finished the drastic decline of manufacturing production. But it did not serve to stimulate a rejuvenation of Chilean manufacturing. In the first six months of 1983, industrial production was only marginally higher than in the corresponding months of 1982 (see Table 11.13). In mid-1983, industrial protection started to be reintroduced on a temporary basis when a 20 per cent surcharge was placed on the import of manufactured goods and a 12 per cent tax on the purchase of foreign exchange. Manufacturers in Chile suddenly had a nominal protection of 42 per cent alongside a realistic exchange rate. In April 1984, the final eclipse of the 'Chicago boys' took place when Pinochet removed them and their sympathisers from all positions of power in the Ministry of the Economy, Treasury, Central Bank and CORFO. The temporary surcharges were declared permanent and a new period of protection for national industry was declared by a new group of civilian and military economists. Industrial production increased by 11 per cent between the first six months of 1983 and 1984. Most of this growth took place in the import-substituting sectors that had suffered ten years of decline.

Such a radical change in the political economy of industrial growth and direction has caused significant changes in the spatial distribution of Chilean manufacturing. In fact, in any analysis of the changing spatial distribution of Chilean industry between 1973 and 1983 the factors of political economy predominate. Such a political economy approach would basically assume that those regions that had previously specialised in import-substituting industry would register significant industrial decline between 1973 and 1983. Meanwhile, those regions that had come to specialise in export-oriented industry would register a certain degree of industrial expansion. Regions in which both import-substituting and export-oriented industry were significant would have both processes operating coincidentally. Examination of regional changes in production and employment in Chilean industry between 1973 and 1980 confirm the operation of such spatial processes (see Tables 11.14 and 11.15).

The previous sections demonstrated that the greatest concentration of import-substituting industry had developed in Santiago. The switch in political economy had as severe an impact on import-substituting industry in the core as in the periphery. The thirteen import-substituting sectors that recorded both production and employment decline nationally had comparable performances in the capital.[51] Export-oriented industry was poorly developed, mainly because Santiago province does not contain one of Chile's major resources. As a result, between 1973 and 1980, industrial employment and production in Santiago declined by 23.1 and 21.3 per cent respectively.

Even more dramatic employment decline occurred in the two peripheral provinces that had significant regional specialisations in import-substituting industry. The provinces of Tarapacá and Concepción recorded respective declines in employment of 40.5 and 33.3 per cent for the period. In the Concepción region, the woollen textile industry of Tomé and the ceramics industry of Penco were virtually eliminated. The glass production of Lirquén and the textile production in Chiguayante were severely depleted. Regional production, however, did not decline due to the statistical impact

Table 11.14: Chile: Changes in Regional Manufacturing Employment, 1973–80

	1973	1980	Percentage Change 1973–80
Tarapacá	10,634	6,332	−40.5
Antofagasta	6,813	9,785	+43.6
Atacama	2,061	2,191	+ 6.3
Coquimbo	2,037	2,553	+25.3
Aconcagua	3,001	3,585	+19.5
Valparaíso	23,187	13,240	−42.9
Santiago	150,729	115,958	−23.1
O'Higgins	5,974	6,171	+ 3.3
Colchagua	868	517	−40.4
Talca/Curicó	3,794	5,299	+39.7
Maule/Linares	1,570	334	−78.7
Ñuble	2,293	2,622	+14.3
Concepción	32,020	21,370	−33.3
Arauco	1,213	905	−25.4
Bío-Bío	4,208	3,419	−18.8
Malleco/Cautín	3,164	3,046	− 3.7
Valdivia	6,232	3,618	−42.0
Osorno	1,861	1,654	−11.1
Llanquihue/Chiloé/Aysén	2,334	3,376	+44.6
Magallanes	979	1,545	+57.8
TOTAL	264,972	207,520	−21.7

Note: Provincial classification of 1973 maintained
Source: R. N. Gwynne, *Production and Spatial Change in Chilean Industry, 1974–1984* (ESRC Research Report, 1985)

of oil refining in San Vicente, the expansion of steel production at Huachipato and fish processing in Talcahuano. In Tarapacá's northern port of Arica, the regional specialisation of electronics and television production virtually terminated and the motor vehicle industry was reduced to the operating of one firm — General Motors. Production declined less than employment in Tarapacá, due to the impact of the booming fish-meal industry — both in Arica itself and Tarapacá's regional capital of Iquique.

Regions more closely associated with Chile's expanding industrial exports recorded significant production increases though less impressive employment increases. Industrial production in the three copper-producing provinces of Antofagasta (Chuqicamata), Atacama (Potrerillos/El Salvador) and O'Higgins (El Teniente) expanded by 136, 97 and 7 per cent respectively — but employment by substantially less. The other major location for copper refining

Table 11.15: Chile: Changes in Regional Manufacturing Employment, 1973–80 (billions of Chilean pesos at 1980 value)

	1973	1980	Percentage Change 1973–80
Tarapacá	20.2	17.9	− 11.4
Antofagasta	23.5	55.5	+ 136.2
Atacama	11.3	22.3	+ 97.4
Coquimbo	2.8	4.9	+ 75.0
Aconcagua	10.6	9.7	− 8.5
Valparaíso	65.5	85.5	+ 30.5
Santiago	259.8	204.5	− 21.3
O'Higgins	21.9	23.4	+ 6.9
Colchagua	4.2	1.3	− 69.1
Talca/Curicó	8.3	12.8	+ 54.2
Maule/Linares	3.5	0.5	− 85.7
Ñuble	2.9	4.0	+ 37.9
Concepción	58.0	63.8	+ 10.0
Arauco	4.8	4.3	− 10.4
Bío-Bío	12.2	7.7	− 36.9
Malleco/Cautín	6.4	5.7	− 10.9
Valdivia	5.6	5.7	+ 1.8
Osorno	5.9	3.4	− 42.4
Llanquihue/Chiloé/Aysén	6.2	3.3	− 46.8
Magallanes	1.2	1.4	+ 16.7
TOTAL	534.8	537.6	+ 0.5

Note: Provincial classification of 1973 maintained
Source: R. N. Gwynne, *Production and Spatial Change in Chilean Industry, 1974–1984* (ESRC Research Report, 1985)

is at Las Ventanas to the north of Valparaíso. The influence of copper and oil refining (at Con Con) in Valparaíso caused a production increase of 30 per cent; but the decline of Valparaíso's import-substituting industry caused a near halving of its manufacturing employment.

The data in Tables 11.14 and 11.15 refer to plants with over fifty employees and, consequently, the opening or closing down of one or two plants in the provinces can have a significant effect on industrial performance. Thus, Coquimbo's expansion was partly linked to the opening up of Goodyear's tyre plant in the regional capital, while the demise of Linares was linked to the closure of a large CORFO plant refining sugar beet and that of Colchagua to the termination of cigarette production in San Fernando.

Outside the copper-producing provinces industrial growth linked to export expansion is more difficult to find. The most dynamic provinces are those of Curicó and Talca, two adjacent provinces in the north of Chile's Central Valley. These provinces (along with Aconcagua) have been at the forefront of the expansion in Chile's fruit-growing industry and Curicó itself has become the fastest growing town of the Central Valley as a result. Further south, expansion in the production of timber, cellulose and paper has had less of an effect on regional indices, as alongside expansion in these sectors, there has been decline in industries oriented to the national market. The decline in Osorno's dairy products industry because of subsidised imports of dairy products from the EEC is one of the most extraordinary cases. But, at the same time, the local industries of Chillán (Ñuble), Temuco (Cautín) and Valdivia have been severely depleted — in particular between 1980 and 1982, when the fixed exchange rate policy caused a massive increase in food imports and further decline in the food industries of the south.

The change in the political economy of Chile between 1973 and 1984 has caused the industrial sector to become a less important part of the national economy. In so doing, a certain degree of industrial decentralisation has taken place. Santiago's proportion of Chile's industrial production diminished from 49 per cent in 1973 to 38 per cent in 1980; its proportion of national value-added was reduced from 48 to 43 per cent in the same period. However, manufacturing employment in Santiago stands as high as ever — at 56 per cent of the national total. As Chilean economic policy reverts to protecting national industry once again, such a major concentration of manufacturing employment and import-substituting firms that 'survived' the previous ten years will be a major attraction for any future growth in industry oriented to the national market.

References

1. J. F. Rippy and J. Pfeiffer, 1948, Notes on the Dawn of Manufacturing in Chile, *Hispanic American Historical Review*, 28, pp. 292–303

2. J. B. Pfeiffer, 1952, Notes on the Heavy Equipment Industry in Chile, 1800–1910, *Hispanic American Historical Review*, 32, pp. 139–44

3. Dirección General de Estadística, 1930, *I Censo Nacional de Industrias, 1928.* (Santiago)

4. J. H. Butler, 1960, *Manufacturing in the Concepción region of Chile.* (National Academy of Sciences and National Research Council, Washington)

5. T. Jeanneret, 1972, El sistema de protección a la industria Chilena, in

O. Muñoz (ed.), *Proceso a la industrialización Chilena*. (Universidad Católica, Santiago)

6. P. T. Ellsworth, 1945, *Chile: an economy in transition*. (Macmillan, New York)
7. M. A. Ballesteros and T. E. Davis, 1963, The growth of output and employment in basic sectors of the Chilean economy, 1908–57, *Economic Development and Cultural Change*, 11, 2, pp. 152–76
8. J. G. Palma, 1979, *Growth and Structure of Chilean Manufacturing Industry from 1830 to 1935*. (Unpublished doctoral thesis, University of Oxford)
9. Ibid.
10. M. A. Ballesteros and T. E. Davis, The growth of output and employment
11. W. P. Glade, 1969, *The Latin American economies*. (Van Nostrand, New York), p. 439.
12. Ibid.
13. J. H. Butler, *Manufacturing in Concepción region*
14. Ibid.
15. M. A. Ballesteros and T. E. Davis, The growth of output and employment
16. M. J. Mamalakis, 1978, *Historical Statistics of Chile: National Accounts, Volume 1*. (Greenwood Press, Westport), pp. 151–3
17. Ibid.
18. W. P. Glade, *The Latin American economies*
19. R. Ffrench-Davis, 1973, *Politicas economicas en Chile, 1952–1970* (Ediciones Nueva Universidad, Santiago), p. 246
20. W. P. Glade, *The Latin American economies*, p. 444
21. Ibid, p. 444
22. S. Bitar, 1972, Hacia la definición de una estrategia industrial, in O. Muñoz (ed.), *Proceso a la industrialización Chilena*. (Universidad Católica, Santiago)
23. M. J. Mamalakis, *Historical Statistics, Volume 1*, p. 202
24. M. J. Mamalakis, 1976, *The Growth and Structure of the Chilean Economy*. (Yale University Press, New Haven), pp. 157–8
25. R. N. Gwynne, 1978, Government planning and the location of the motor vehicle industry in Chile, *Tijdschrift voor economische en sociale geografie*, 69, pp. 130–40
26. Ibid.
27. O. Muñoz, 1972, Crecimiento industrial, estructura del consumo y distribución del ingreso, in O. Muñoz (ed.), *Proceso a la industrialización Chilena*. (Universidad Católica, Santiago), p. 23
28. L. Pacheco, 1972, La inversión extranjera y las corporaciones internacionales en el desarrollo industrial chileno, in O. Muñoz (ed.), *Proceso a la industrialización Chilena*. (Universidad Católica, Santiago), pp. 106–55
29. Ibid., p. 120
30. M. J. Mamalakis, *Historical Statistics, Volume 1*, p. 203
31. R. N. Gwynne, 1976, Economic Development and Structural Change: the Chilean Case, 1970–1973, *Occasional Publication No. 2*, Dept. of Geography, University of Birmingham
32. Instituto de Economía, 1972, *La Economía Chilena, 1971*. (Universidad de Chile, Santiago)
33. F. G. Gil, 1966, *The political system of Chile*. (Houghton Mifflin, Boston)
34. Instituto de Economía, 1973, *La Economía Chilena, 1972*. (Universidad de Chile, Santiago)
35. El Mercurio, 10 July 1973
36. Instituto Nacional de Estadisticas, *Indices de Producción Industrial* (August 1973)
37. R. N. Gwynne, 1978, Industrial development in the periphery: the motor

vehicle industry in Chile, *Bulletin of the Society for Latin American Studies*, 29, pp. 47–69

38. R. N. Gwynne, 1980, Import substitution and the decentralisation of industry in less developed countries: the television industry in Chile, 1962–1974, *Occasional Publication No. 12*, Dept. of Geography, University of Birmingham

39. Ibid.

40. H. Blakemore, 1981, Chile: the real revolution, *Bank of London and South America Review*, 15, 4, pp. 180–8

41. L. Gotuzzo, 1979, Three years to end protection: new tariff schedule under study, in J. C. Mendez (ed.), *Chilean Economic Policy*. (Central Bank of Chile, Santiago), pp. 75–82

42. Instituto Nacional de Estadisticas

43. Industrial Surveys of 1973 and 1976

44. Banco Central de Chile, 1981, *Precios y Cantidades Fisicas de Principales Productos de Exportación e Importación, Diciembre 1980*. (Banco Central Santiago)

45. D. Massey and R. Meegan, 1982, *The Anatomy of Job Loss*. (Methuen, London)

46. H. Blakemore, Chile: the real revolution

47. E. Errazuriz, 1984, La desindustrialización del país, *Mensaje*, 33, 326, pp. 41–5

48. Banco Central, 1982, *Series de Comercio Exterior, 1970–1981*. (Banco Central, Santiago)

49. R. N. Gwynne, 1985, Production and Spatial Change in Chilean Industry, 1974–1984 (ESRC Research Report)

50. El Mercurio, December 14th, 1982

51. R. N. Gwynne, Production and Spatial Change in Chilean industry 1974–1984

12 CONCLUSION: THE IMPORTANCE OF INDUSTRIAL GROWTH

This book has attempted to develop two basic arguments. First, it has tried to show the importance of a process of industrialisation for all Latin American countries and the necessity for intermediately-developed countries to plan actively for the growth of the manufacturing sector. Secondly, it has been argued that the spatial development of industrialisation within Latin American countries has had a significant impact on the spatial incidence of urban and regional growth.

In the first part of the book, the position of Latin American countries within the world economy was explored. In terms of post-war economic growth, Latin American countries that had industrialised had recorded higher growth rates than countries that had not. In particular, the ability of Latin American countries to confront the debt crisis and recession of the 1980s has been closely linked to the level of industrialisation. Those countries with mature industrial structures have been able to relieve the considerable external constraints on their economies to a much greater extent than those countries that have less mature industrial structures or have ignored the performance of the manufacturing sector. One country to ignore the manufacturing sector during the 1970s and early 1980s was Chile and the Chilean case study in the final chapter demonstrated the serious implications that the consequent industrial decline now has for the future development of the national economy.

The gravest implications of Chile's industrial decline would appear to be in terms of its international trading potential and its relationship with the world economy. In order to understand the scale of the present problem, one must make reference to the massive inflow of foreign capital into Chile between 1977 and 1981, and the consequent creation of a highly indebted country. In terms of the relationship between *per capita* debt and *per capita* income, Chile is now the fourth most indebted country in Latin America (see Chapter One) with *per capita* debt fast approaching the level of *per capita* income. With high world interest rates forecast for the rest of the 1980s, such indebtedness provides a major external constraint on development. At present, Chile cannot even pay the annual interest payments on the debt. According to an agreement with the International Monetary Fund in 1984, Chile had to pay approximately 1,800 million dollars in interest payments on its total debt for that year; about 1,000 million dollars was to come from a surplus on the balance of trade. By the end of the year, however, the balance of trade surplus was only 176 million dollars — 17.6 per cent of the forecast surplus and only 9.8 per cent of total annual interest payments.

The critical point about such massive debt is that the only way to reduce its constraining influence on the national economy is to increase exports. The policy of boosting exports has been followed by both Brazil and Mexico during the debt crisis and both countries have managed to reduce the external constraints on their economies as a result. However, their success in boosting exports has relied heavily on the expansion of manufacturing exports, particularly to the United States in the period of the strong dollar (1983–1985). In 1984, manufactured exports came to represent 54 per cent of total Brazilian exports of 27 billion dollars. Between 1983 and 1984, exports of Brazilian semi-manufactured goods rose by 63 and 32 per cent respectively; the value of primary exports rose by only 14 per cent, mainly due to increased production rather than increased prices.

Unfortunately, Chile has been unable to adopt such an export policy due to its depleted industrial structure. It has had to rely almost exclusively on the export of primary products. As a result, the value of its exports have actually fallen — from 3,906 million dollars in 1981 to 3,657 million dollars in 1984. The volume of primary product exports often increased but prices declined. An average weighted decline of 12.2 per cent took place in the prices of Chile's twenty leading exports during the first three years of the debt crisis, 1981–83. Despite particular blame being placed on low copper prices by the Chilean government, copper prices declined by less than 12 per cent between 1981 and 1983.

This lack of a manufacturing base in Chile has meant that it has not been possible to boost exports during the 1981–85 debt crisis, unlike Brazil and Mexico. It has been unable, therefore, to reduce the external constraints of large debts, annual interest payments and shortage of foreign and domestic capital. As was pointed out in Chapter Eleven, a more favourable attitude towards the Chilean manufacturing sector occurred in 1984. However, this late attempt to boost manufacturing in Chile had a negative short-term effect because it understandably caused the import of capital goods to increase — by a total of 54 per cent in 1984. The manufacturing sector, which had been squeezed for the previous three years, needed to import new capital equipment before productive expansion could take place. Thus, a further external constraint has been added due to past industrial decline. If manufacturing production is to expand in the long-term (and manufacturing exports increase), there is a short-term problem of increased imports, which acts to reduce the trade surplus. The legacy of Chilean deindustrialisation is therefore an uncomfortable conflict between two very necessary short-term aims — debt repayment and economic growth. The Chilean example clearly shows the importance of nurturing manufacturing industry. A dynamic manufacturing sector tends to reduce the external constraints of the world economy on national development, particularly in times of recession when primary product prices are at their lowest.

Nevertheless, the analysis in Chapter Two demonstrated that within the framework of the traditional industrial policy of Latin America, import substitution, small-scale and high-cost industries became characteristic of small countries. The countries that benefited most from import substitution industrialisation were the larger countries. Furthermore, these countries found it easier to switch the emphasis of policy to export promotion — as they already possessed large-scale industrial firms and plants that could compete internationally with backing from their respective governments. Smaller countries must begin to specialise in certain manufacturing sectors at an earlier stage in their economic development than larger countries. As with the recent record of the larger countries, governments must play an important role in promoting manufacturing exports in certain key sectors.

The greater success of policies of import substitution in larger countries caused a more mature development of industry and created industrial structures comparable with those of developed countries. Brazil has the ability to manufacture in all major sectors — consumer durables and non-durables, intermediate and capital goods. In order to have a viable capital goods sector, it has been necessary to develop the ability not only to adapt and improve imported technologies but also to create new technologies, as with the Brazilian Alcohol Programme. Smaller Latin American countries with less mature industrial structures have been unable to develop new technologies and have had a poor record in adapting and improving imported technologies. As a result, they are tending to depend increasingly on adapted technologies imported from the larger countries of Latin America.

Brazil has been able to develop a mature industrial structure and a refreshing degree of technological independence due to an interesting institutional alliance between state firms, multinational corporations and national companies. To a certain extent, each has developed a different role. State firms have been dominant in the provision of infrastructure, the production of basic intermediate goods and in those sectors considered as being of national and strategic interest. Multinational corporations have been increasingly dominant in those technologically-innovative sectors considered to be outside areas of strategic importance. Meanwhile, national companies are much more diverse in character, but are certainly prominent in the more labour-intensive and traditional sectors of the economy.

Considerable criticism has often been levelled at the involvement of multinational corporations in Latin American manufacturing. Certainly the fact that multinational enterprise is playing an important role in the more technologically innovative sectors is potentially serious for the future stability of industrial growth. However, often the comparison between multinational enterprise and national companies shows the former to consist of specialised and stable manufacturing entities needing to maintain a solid worldwide reputation in order to further future business projects. The record of national conglomerates towards manufacturing industry can sometimes seem irresponsible in comparison. Returning to the Chilean case study, it was demonstrated how most of the national conglomerates that bought up manufacturing companies in the 1970s did so as a sideline to their major interests in banking and finance. When the 1982 Chilean devaluation suddenly caused the virtual collapse of their financial empires (with dollar debts greater than dollar assets), the manufacturing subsidiaries began to be milked of all the profits they made in order to pay off the huge debts that corresponded to the conglomerate as a whole. As a result, little new investment has occurred in the manufacturing companies that belong to the major Chilean conglomerates since 1982.

Industrial success in Latin America has been matched by its spatial concentration. The largest cities of the continent, São

Paulo and Mexico City, are precisely those with the largest and most dynamic industrial complexes. It must be remembered that industrial expansion in the twentieth century has been implanted upon a historical urban pattern that was distinctly primate in nature, the inheritance, in Spanish America at least, of hierarchical colonial administration, regional independence movements and the booming export economies of the nineteenth century. Industrialisation, however, has exacerbated this historical concentration of population and economy. Government attempts to decentralise industry have often met with failure. Even in those cases that have been successful, little impact has been made on the overall spatial concentration of industry. The only major decentralising phenomenon, that of short-distance decentralisation from the primate city, can best be seen not as a successful form of industrial deconcentration but as a further strengthening and intensifying of the process of centralisation.

In this way, income disparities between regions *with* and regions *without* mature industrial structures have tended to increase rather than decrease. Streams of migration from the poorer regions to the richer regions of individual countries have not noticeably affected the incidence of regional disparities. Hirschman's prediction of the equalising effect of migratory streams from poorer to richer regions in the 1950s has therefore not been fulfilled. The exaggerated spatial concentration of industry has caused disparities in both productivity and income between regions to continue to increase. The bell-shaped curve of Alonso (see Chapter Ten) can still not be applied to the evolution of regional disparities in Latin America. In short, Latin America is developing a radically different spatial pattern to any other continent as it industrialises.

However, industrialisation has not been able to provide the majority of employment in any urban area. The proportion of workers in the manufacturing sector ranges from 20 to 35 per cent in the larger cities of Latin America. Generally speaking, the larger the city, the higher the proportion of industrial workers as a percentage of the total work force. As a result of the relatively low proportions of urban workers engaged in manufacturing activities, there is a large small-scale sector employed in tertiary activities. Nevertheless, the large industrial city has shown and continues to show a great flexibility in providing work for the migrant masses — even in times of recession. For example, in 1984, a year of some recovery in the Chilean economy after two years of severe decline,

155,000 new jobs (net) were created. The spatial distribution of job creation demonstrated the greater flexibility of the metropolitan economy of Santiago. Santiago now accounts for as much as 40 per cent of Chile's work force. But in 1984 it accounted for as much as 60 per cent of the new jobs created.

The process of industrialisation and its spatial concentration has undoubtedly fuelled migration. However, there is a lack of coincidence between the opportunities of industrial employment and urban population growth. Consequently, an urban employment structure with a large small-scale sector has developed. Furthermore, with four years of recession for many Latin American countries (1981–85) and consequent reductions in employment, urban society has been characterised by increasing numbers of marginal populations. For example, in Santiago, unemployment increased from 8 per cent in mid-1981 to 32 per cent by the end of 1983. But in the squatter settlements of Santiago, such as La Victoria, unemployment figures of 80 per cent have been estimated. Nevertheless, for the migrant, the large industrial city is not so much a source of manufacturing employment as a symbol of material advance. The manufacturing production of the industrial city may give little employment in relation to the totality of migrants, but the city still has an image closely linked to that of material progress.

BIBLIOGRAPHY

Alonso, W. 1975, in Friedmann, J. and Alonso, W. *Regional Policy/Readings in Theory and Applications.* (MIT, Cambridge, Mass.), pp. 64–96

——— 1980, Five bell shapes in development, *Papers of the Regional Science Association*, 45, pp. 5-16

Assadourian, Carlos S. 1973, Sobre un elemento de la economía colonial: producción y circulación de mercancías al interior de un conjunto regional, *Revista Latinamericana de Estudios Urbano Regionales (EURE)* 3 December, pp. 135–81

Avery, W. P. and Cochrane, J. D. 1972, Subregional Integration in Latin America: The Andean Common Market, *Journal of Common Market Studies*, Vol. XI, No. 2, p. 85

Baer, W. 1962, The economics of Prebisch and ECLA, *Economic Development and Cultural Change*, 10, pp. 169–82

——— 1965, *Industrialization and Economic Development in Brazil.* (Irwin, Homewood, Illinois)

——— 1969, *The Development of the Brazilian Steel Industry.* (Vanderbilt University Press, Tennessee)

———, Kerstenetsky, I. and Villela, A. 1973, The changing role of the State in the Brazilian economy, *World Development*, VI, 11, pp. 23–34

Baerresen, D. 1971, *The Border Industrialization Program of Mexico.* (D. C. Heath & Co., Lexington, Mass.)

Balán, J., Browning, H. and Jelin, E. 1973, *Man in a developing society.* (University of Texas Press, Austin and London)

Balassa, B. 1971, *The Structure of Protection in Developing Countries.* (John Hopkins Press, Baltimore)

Ballesteros, M. A. and Davis, T. E. 1963, The Growth of Output and Employment in Basic Sectors of the Chilean Economy, 1908–57, *Economic Development and Cultural Change*, 11(2). pp. 152–76

Banco Central de Chile, 1981, *Precios y Cantidades Físicas de Principales Productos de Exportación e Importación, Diciembre 1980.* (Banco Central, Santiago)

——— 1982, *Series de Comercio Exterior, 1970–1981.* (Banco Central, Santiago)

Barkin, D. and King, T. 1970, *Regional Economic Development: The River Basin Approach in Mexico.* (Cambridge University Press, Cambridge)

Bergsman, J. 1970, *Brazil, industrialisation and trade policies.* (Oxford University Press, London)

Berry, B. J. L. 1969, Relationships between regional economic development and the urban system; the case of Chile. *Tijdschrift voor Economische en Sociale Geografie*, 60(5), pp. 283–307

Bitar, S. 1972, Hacia la definición de una estrategia industrial, in Muñoz, O. (ed.) *Proceso a la industrialización Chilena.* (Universidad Católica, Santiago)

Blackbourn, A. 1974, The Spatial Behaviour of American Firms in Western Europe, in Hamilton, F. E. I. (ed.) *Spatial Perspectives on Industrial Organisation and Decision-making.* (J. Wiley & Sons, London), pp. 245–264

Blakemore, H. 1971, Chile, in Blakemore, H. and Smith, C. T. (eds) *Latin America: Geographical Perspectives.* (Methuen, London)

——— 1981, Chile: The Real Revolution, *Bank of London and South America Review*, 15, IV, pp. 180–8

——— and Smith, C. T. (eds) 1971, *Latin America: Geographical Perspectives.* (Methuen, London)

Boisier, S. 1974, Localización tamaño urbano y productividad industrial: un caso de estudio de Brasil, *Revista Latinamericana de Estudios Urbano Regionales*, 3(9), pp. 57–78

Bottomley, A. 1965, Imperfect Competition in the Industrialization of Ecuador, *Inter-American Economic Affairs*, 19 January, pp. 83–94

Boudeville, J. R. 1966, *Problems of Regional Economic Planning.* (Edinburgh University Press, Edinburgh)

Brading, David A. 1971, *Mines and merchants in Bourbon Mexico, 1763–1810.* (Cambridge University Press, Cambridge)

Bromley, R. (ed.) 1979, *The urban informal sector.* (Pergamon, Oxford)

Brookfield, H. C. 1975, *Interdependent Development.* (Methuen, London)

Butler, J. H. 1960, *Manufacturing in the Concepción Region of Chile: present position and prospects for future development.* (National Academy of Sciences and National Research Council, Washington D.C.)

Butterworth, D. and Chance, J. K. 1981, *Latin American Urbanisation.* (Cambridge University Press, Cambridge)

Cardoso, F. H. and Faletto, E. 1979, *Dependency and Development in Latin America.* (University of California Press, Berkeley)

Chenery, H. B. 1977, Transitional Growth and World Industrialisation, in Ohlin, B. et al. (eds) *The International Allocation of Economic Activity* (Macmillan, London and Basingstoke)

Cody, J., Hughes, H. and Wall, D. (eds) 1980, *Policies for industrial progress in developing countries.* (Oxford University Press, Oxford)

Cohen, R. B. 1981, The new international division of labor, multinational corporations and urban hierarchy, in Dean, M. and Scott, A. J. *Urbanisation and Planning in Capitalist Society.* (Methuen, London), pp. 287–315

Congdon, T. G. 1982, "Apertura" Policies in the Cone of Latin America, *The World Economy*, 5(2), pp. 133–48

Cornelius, W. A. and Kemper, R. V. (eds) 1978, *Latin American Urban Research, 6,* (Sage Publications, Beverley Hills and London)

——— and Trueblood, F. M. (eds) 1974, *Latin American Urban Research, 4,* (Sage Publications, Beverley Hills and London)

———, ——— 1975, *Latin American Urban Research, 5,* (Sage Publications, Beverley Hills and London)

Cunningham, S. M. 1981, Multinational Enterprises in Brazil: Locational Patterns and Implications for Regional Development, *Professional Geographer*, Vol. 33, No. 1, p. 48

——— 1982, Industrial Estates as a Planning Tool: Recent Experiences in Rio de Janeiro and Minas Gerias States, Brazil, *Third World Planning Review*, 4(1), pp. 44–60

Dahse, F. 1979, *El mapa de la extrema riqueza*, Editorial Aconcagua, Santiago

Dean, W. 1969, *The Industrialization of São Paulo, 1880–1945.* (University of Texas, Austin)

Dicken, P. 1971, Some aspects of the decision-making behaviour of business organisations, *Economic Geography*, 47, pp. 426–38

Dickenson, J. P. 1967, The Iron and Steel Industry in Minas Gerais, Brazil, 1695–1965, in Steel, R. W. and Lawton, R. (eds) *Liverpool Essays in Geography* (Longman, London)

——— 1970, Industrial Estates in Brazil, *Geography*, 55, July, pp. 326–9

——— 1978, Industrial Estates and the Location of Industry in Brazil, *Bank of London and South America Review*, 12, 4, pp. 176–184

——— 1978, *Brazil.* (Dawson, Folkestone)

Dillman, C. D. 1970, Recent Developments in Mexico's National Border Programme, *The Professional Geographer*, 22, September, pp. 243–7

——— 1970, Urban Growth along Mexico's Northern Border and the Mexican National Border Programme, *Journal of Developing Areas*, 4, July, pp. 487–508

Dirección General de Estadistica 1930, *I Censo Nacional de Industrias (1928)* Santiago, Chile

Dirección General de Estadistica y Censos Nacionales 1976, *IV Encuesta Industrial 1974: Resultados Regionales* (Ministerio de Fomento, Caracas)

——— 1977, *V Encuesta Industrial 1975: Resultados Regionales* (Ministerio de Fomento, Caracas)

Dirección General Sectorial de Industrias 1977, *Política de Desconcentración Industrial* (Ministerio de Fomento, Caracas)

Ellsworth, P. T. 1945, *Chile: An Economy in Transition*. (Macmillan, New York)

Errázuriz, E. 1984, La desindustrialización del país, *Mensaje*, 33, 326, pp. 41–45

Faria, V. E. 1978, *Occupational marginality, employment and poverty in urban Brazil*. (Unpublished doctoral dissertation, Harvard University)

Felix, D. 1977, The technological factor in socio-economic dualism: toward an economy-of-scale paradigm for development theory, *Economic Development and Cultural Change*, 25 (supp), pp. 180–211

Fernández, M. P. 1981, The U.S.-Mexico Border: Recent Publications and the State of Current Research, *Latin American Research Review*, 16, 3, pp. 250–267

Ffrench-Davis, R. 1973, *Políticas Económicas en Chile, 1952–1970*, Centro de Estudios de Planificación Nacional (CEPLAN), Ediciones Nueva Universidad, Universidad Católica de Chile, Santiago

——— 1982, Foreign Trade, Industrialisation and Development Policies, in Ffrench-Davis, R. and Tironi, E. *Latin America and the New International Economic Order*. (Macmillan, London), pp. 157–185

——— and Tironi, E. (eds) 1982, *Latin America and the New International Economic Order*. (Macmillan, London and Basingstoke)

Frank, A. G. 1971, *Capitalism and Underdevelopment in Latin America*. (Penguin, Harmondsworth)

Franko, L. G. 1976, *The European Multinationals*. (Harper & Row, London)

Friedmann, J. 1966, *Regional Development Policy: A Case Study of Venezuela*. (MIT Press, Cambridge, Mass.)

——— 1969, The Role of Cities in National Development, *American Behavioural Scientist*, Vol. 12, No. 5

——— 1970, Urban-Regional Policies for National Development in Chile, in Rabinovitz, F. and Trueblood, F. (eds) *Latin American Urban Research*. (Sage, Beverley Hills), Vol. 1, p. 217

——— 1973, *Urbanisation, Planning and National Development*. (Sage Publications, Beverley Hills)

——— and Alonso, W. 1975, *Regional Policy: Readings in Theory and Applications*. (MIT Press, Cambridge, Mass.)

——— and Lackington, T. 1967, Hyperurbanisation and National Development in Chile: Some Hypotheses, *Urban Affairs Quarterly*, 11 June, pp. 3–29

——— and Weaver, C. 1979, *Territory and Function: The evaluation of regional planning*. (Edward Arnold, London)

——— and Wulff, R. 1975, *The Urban Transition: comparative studies of newly industrialising societies*. (Edward Arnold, London)

Frisbie, W. Parker, 1976, *The scale and growth of world urbanisation*. (Population Research Center, University of Texas at Austin)

Fox, Robert W. 1975, *Urban population growth trends in Latin America*. (Inter-American Development Bank, Washington, D.C.)

Furtado, C. 1970, *Economic Development of Latin America: A Survey from Colonial Times to the Cuban Revolution*. (Cambridge Univeristy Press, London)
Garza, G. and Schteingart, M. 1978, Mexico City: The Emerging Metropolis, *Latin American Urban Research*, 6, pp. 51–86
Geertz, C. 1963, *Agricultural involution: the process of ecological change in Indonesia*. (University of California Press, Berkeley)
Geisse, G. and Hardoy, J. E. (eds) 1972, *Latin American Research, Volume 2, Regional and Urban Development Policies — A Latin American Perspective*. (Sage Publications, Beverley Hills)
Gereffi, G. and Evans, P. 1981, Transnational Corporations, Dependent Development, and State Policy in the Semiperiphery: A Comparison of Brazil and Mexico, *Latin American Research Review*, XVI, 3, pp. 31–64
Gil, F. G. 1966, *The Political System of Chile*. (Houghton Mifflin, Boston)
Gilbert, A. G. 1974, Industrial location theory: its relevance to an industrialising nation, in Hoyle, B. S. (ed.) *Spatial aspects of development*. (John Wiley, London), pp. 271–290
——— 1974b, *Latin American development: a geographical perspective*. (Penguin, Books, Harmondsworth)
——— 1976, *Development planning and spatial structure*. (John Wiley, London)
——— 1976, The arguments for very large cities reconsidered, *Urban Studies*, 13, pp. 27–34
——— 1977, The argument for very large cities reconsidered: a reply, *Urban Studies*, 14, pp. 225–7
——— and Goodman, D. 1976, Regional income disparities and economic development: a critique, in Gilbert, A. (ed.) *Development planning and spatial structure*. (John Wiley, London)
Glade, William 1969, *The Latin American economies: a study of their institutional evolution*. (Van Nostrand, New York)
Gómez, A. 1974, El proceso de localización industrial en Chile: análisis y políticas, *Revista Latinamericana de Estudios Urbano Regionales*, 111, 9, pp. 9–56
Goodman, D. E. 1972, Industrial Development in the Brazilian North-East: An Interim Assessment of the Tax Credit Scheme of Article 34/18, in Roett, R. J. A. (ed.) *Brazil in the Sixties*. (Vanderbilt Press, Nashville)
——— 1976, The Brazilian Economic "Miracle" and Regional Policy: Some Evidence from the Urban Northeast, *Journal of Latin American Studies*, 8, 1, pp. 1–27
Gotuzzo, L. 1979, Three years to end protection: new tariff schedule under study, in Mendez, J. C. (ed.) *Chilean Economic Policy*. (Central Bank of Chile, Santiago), pp. 75–82.
Griffin, K. B. 1969, *Underdevelopment in Spanish America: An Interpretation*. (Allen and Unwin, London)
Griffith-Jones, S. 1982, Transnational Finance and Latin American National Development, *IDS Discussion Paper 175*, Sussex
Grunwald, J. 1970, Some Reflections on Latin American Industrialization Policy, *Journal of Political Economy*, 78, July-August, pp. 826–56
Gunderson, G. S. 1979, The worldwide corporation — an economic catalyst. (IGU Commission on Industrial Systems, Rotterdam)
Gwynne, R. N. 1976, *Economic Development and Structural Change: The Chilean Case, 1970–1973*. (Department of Geography, University of Birmingham, Occasional Publication No. 2)
——— 1978, *Industrial Decentralisation in Chile: The Case of Arica*. (Unpublished doctoral dissertation, Liverpool University)
——— 1978b, The Motor-Vehicle Industry in Latin America, *Bank of London and South America Review*, 12(9), pp. 426–71

——— 1978c, Government Planning and the Location of the Motor Vehicle Industry in Chile, *Tijdschrift voor Econ. en Soc. Geografie*, 69, No. 3

——— 1978d, City Size and Retail Prices in Less Developed Countries, *Area*, 10(2), pp.136–40

——— 1978e, Industrial Development in the Periphery: The Motor Vehicle Industry in Chile, *Bulletin of the Society for Latin American Studies*, No. 29, November, p. 47

——— 1979, The Venezuelan automobile industry, *Business Venezuela*, 64, pp. 24–28

——— 1979b, The plight of productivity, *Business Venezuela*, 63, pp. 27–30

——— 1979c, Oligopolistic reaction, *Area*, 11(4), pp. 315–19

——— 1980, The Andean Group Automobile Programme: an interim assessment, *Bank of London and South America Review*, 14, No. 11, p. 160

——— 1980b, *Import Substitution and the Decentralisation of Industry in Less Developed Countries: The Television Industry in Chile, 1962–1974*. (Department of Geography, University of Birmingham Occasional Publications No. 12)

——— 1981, Chile is a case-study in monetarism, *Geographical Magazine*, 53(6), pp. 361–71

——— 1982, Location Theory and the Centralization of Industry in Latin America, *Tijdschrift voor Economische en Sociale Geografie*, 73(2), pp. 80–93

——— and Cunningham, S. 1983, The greatest debtors in the world, *Geographical Magazine*, 55, 11, pp. 569–72

——— 1983, When trade stops being the engine of growth, *Geographical Magazine*, 55, 10, pp. 503–7

Halperin, D. T. 1970, *Historia contemporánea de América Latina*. (Alianza Editorial, Madrid)

Hamilton, F. E. I. 1985, Multinational enterprise: spectre or spearhead, in Watts, D., Drakakis-Smith, D. and Dixon, C. *Multinational Companies and the Third World*. (Croom Helm, London, forthcoming)

Hardoy, J. (ed.) 1975, *Urbanization in Latin America: approaches and issues*. (Anchor, New York)

Hauser, Philip M. (ed.) 1961, *Urbanisation in Latin America*. (UNESCO, Paris)

Hilhorst, J. G. M. 1980, On unresolved issues in regional development thinking. *Institute of Social Studies, The Hague, Occasional Papers* No. 81

——— 1981b, Territory vs function: a new paradigm? *Institute of Social Studies, The Hague, Occasional Papers* No. 89

Hirschman, A. O. 1958, *The Strategy of Economic Development*. (Yale University Press, New Haven)

——— 1968, The Political Economy of Import-Substituting Industrialization in Latin-American Countries, *Quarterly Journal of Economics*, 82, pp. 1–32

——— 1968b, Industrial Development in the Brazilian Northeast and the Tax Credit Scheme of Article 34/18, *Journal of Development Studies*, 5 October, pp. 1–28

Hojman, D. 1981, The Andean Pact: Failure of a Model of Economic Integration, *Journal of Common Market Studies*, XX, No. 2, pp. 139–60

Humphrey, J. 1982, *Capitalist control and workers' struggle in the Brazilian auto industry*. (Princeton University Press, Princeton)

Hunter, J. M. and Foley, J. W. 1975, *Economic Problems of Latin America*. (Houghton Mifflin, Boston, Mass.)

Hymer, S. 1975, The Multinational Corporation and the Law of Uneven Development, in Radice, H. (ed.) *International Firms and Modern Imperialism*. (Penguin, Harmondsworth), pp. 37–62

Instituto de Costos 1969, *Estudio sobre la industria automotriz*. Santiago

Instituto de Economía 1972, *La Economía Chilena, 1971*. (Universidad de Chile,

Santiago)
——— 1973, *La Economía Chilena, 1972*. (Universidad de Chile, Santiago)
International Labour Office (ILO) 1972, *Employment, Incomes and Equality: A Strategy for Increasing Productivity Employment in Kenya*. (ILO, Geneva)
Jeanneret, T. 1972, El sistema de protección a la industria chilena, in Muñoz, O. (ed.) *Proceso a la Industrialización Chilena*. (Universidad Católica, Santiago)
Jenkins, R. O. 1977, *Dependent Industrialization in Latin America: The Automotive Industry in Argentina, Chile and Mexico*. (Praeger Publishers, New York)
——— 1978, Manufactured Exports — Development Strategy or Internationalisation of Capital, *Bulletin of the Society for Latin American Studies*, 28, pp. 64–82
Johnson, L. L. 1967, Problems of Import Substitution: the Chilean Automobile Industry, *Economic Development and Cultural Change*, 15, pp. 202–16
Karlsson, W. 1975, *Manufacturing in Venezuela*. (Almquist and Wiksell, Stockholm)
Katz, J. 1982, Technological Change and Development in Latin America, in Ffrench-Davis, R. and Tironi, E. (eds) *Latin America and the NIEO* (Macmillan, London), pp. 192–211
Kennelly, R. A. 1955, The Location of the Mexican Steel Industry, *Revista Geográfica* 15, pp. 109–29; 16, pp. 199–213; 17, pp. 60–77
Kirkpatrick, C. H. and Nixson, F. I. 1983, *The industrialisation of less developed countries*. (Manchester University Press, Manchester)
Kitching, G. 1982, *Development and Underdevelopment in Historical Perspective*. (Methuen, London)
Knickerbocker, F. T. 1973, *Oligopolistic Reaction and the Multinational Enterprise*. (Harvard University Press, Cambridge, Mass.)
Kowarick, L. 1975, *Capitalismo e marginalidade na América Latina*. (Paz e Terra, Rio de Janeiro)
Kronish, R. and Mericle, K. S. 1984, *The Political Economy of the Latin American Motor Vehicle Industry*. (MIT Press, Cambridge, Mass.)
Kuklinski, A. 1972, *Growth poles and growth centres in regional planning*. (Mouton, Paris)
Kuznets, S. 1973, Modern economic growth: findings and reflections, *American Economic Review*, 63, 3, pp. 247–58
Lagos Escobar, R. 1960, *La Industria en Chile: Antecedentes Estructurales*. (Universidad de Chile, Santiago)
Laite, A. J. 1981, *Industrial development and migrant labour*. (Manchester University Press, Manchester)
Lavell, A. M. 1972, Regional Industrialization in Mexico: Some Policy Considerations, *Regional Studies*, 6 September, pp. 343–62
Lensky, A. S. 1965, Some Generalisations Concerning Primate Cities, *Annals of the Association of American Geographers*, Vol. 55, pp. 506–13
Levi de López, S. 1980, Industrial Development in Mexico, *Bulletin of the Society for Latin American Studies*, 32, pp. 6–24
Lewis, W. A. 1955, *The Theory of Economic Growth*. (George Allen & Unwin, London)
Lipton, M. 1977, *Why Poor People Stay Poor: Urban Bias in World Development*. (Temple Smith, London)
Little, I., Scitovsky, T. and Scott, M. 1970, *Industry and Trade in Some Developing Countries: A Comparative Study*. (Oxford University Press, Oxford)
Lloyd, P. E. and Dicken, P. 1977, *Location in Space: A Theoretical Approach to Economic Geography*. (Harper & Row, London)
McGreevey, W. P. 1971, A statistical analysis of primacy and log normality in the

size distributions of Latin American cities, 1750–1960, in Morse, R. (ed.) *The urban development of Latin America, 1750–1920*. (Center for Latin American Studies, Stanford University)

Malloy, J. M. (ed.) 1977, *Authoritarianism and corporatism in Latin America*. (University of Pittsburgh Press, Pittsburgh)

Mamalakis, M. J. 1976, *The Growth and Structure of the Chilean Economy: from Independence to Allende*. (Yale University Press, New Haven)

——— 1978, *Historical Statistics of Chile: National Accounts*. (Greenwood Press, Westport, Conn.)

——— 1980, *Historical Statistics of Chile: Demography and Labour Force*. (Greenwood Press, Westport, Conn.)

Martner, G. (ed.) 1971, *El Pensamiento económica del gobierno de Allende*. (Editorial Universitaria, Santiago)

Massey, D. and Meegan, R. 1982, *The Anatomy of Job Loss*. (Methuen, London)

Middlebrook, K. J. 1978, Regional Organizations and Andean Economic Integration, 1969–1975, *Journal of Common Market Studies*, 17, 1, pp. 62–82

Miller, R. 1976, Railways and economic development in central Peru, 1880–1930, in Miller, R., Smith, C. T. and Fisher, J. (eds). *Social and economic change in modern Peru*. (Centre for Latin American Studies, University of Liverpool), pp. 27–52

Ministerio de Energía y Minas, 1983, *Petróleo y Otros Datos Estadísticos 1982*, Caracas

Morawetz, David 1974, *The Andean Group: A Case Study in Economic Integration among Developing Countries*. (MIT Press, Cambridge, Mass.)

Morley, S. A. 1983, *Labor Markets and Inequitable Growth: The Case of Authoritarian Capitalism in Brazil*. (Cambridge University Press, Cambridge)

———, and Smith, G. W. 1977, Limited Search and the Technology Choices of Multinational Firms in Brazil, *The Quarterly Journal of Economics*, 91, pp. 263–87

———, ——— 1977b, The Choice of Technology: Multinational Firms in Brazil, *Economic Development and Cultural Change*, 25, pp. 239–64

Morris, A. S. 1979, *South America*. (Hodder & Stoughton, London)

Morris, A. 1981, *Latin America: economic development and regional differentiation*. (Hutchinson, London)

Morse, R. M. 1965, Recent Research on Latin American Urbanisation: A Selective Summary with Commentary, *Latin American Research Review*, Vol. 1, No. 1, pp. 35–74

——— 1970, São Paulo: Case Study of a Latin American Metropolis, in Rabinovitz, F. and Trueblood, F. (eds.) *Latin American Urban Research Vol. 1* (Sage, Beverley Hills), p. 151

——— 1971, Trends and Issues in Latin-American Urban Research, 1965–1970, *Latin American Research Review*, Vol. 6, No. 1, pp. 3–52

——— (ed.) 1971b, *The urban development of Latin America, 1750–1920*. (Center for Latin American Studies, Stanford University)

——— 1974, Trends and patterns of Latin American urbanisation, 1750–1920, *Comparative Studies in Society and History*, 16, 4 (Sept.), pp. 416–47

——— 1975, The development of urban systems in the Americas in the nineteenth century. *Journal of Interamerican Studies and World Affairs*, 17, 1, pp. 4–25

Mountjoy, A. B. 1963, *Industrialization and Under-developed Countries*. (Hutchinson, London)

Muñoz, O. (ed.) 1972, *Proceso a la Industrialización Chilena*. (Universidad Católica de Chile, Santiago)

——— 1972b, Crecimiento industrial, estructura del consumo y distribución del ingreso, in Muñoz, O. (ed.) *Proceso a la Industrialización Chilena*. (Universidad

Católica de Chile, Santiago)
Myrdal, G. 1957, *Economic Theory and Underdeveloped Regions*. (Duckworth, London)
Odell, P. R. and Preston, D. A. 1973, *Economies and Societies in Latin America: A Geographical Interpretation*. (John Wiley, London)
Oficina de Planificación Nacional (ODEPLAN) 1968, *Política de desarrollo*. (Santiago)
Onyemelukwe, J. O. C. 1974, Industrial location in Nigeria, in Hamilton, F. E. Ian (ed.) *Spatial Perspectives on Industrial Organisations and Decision-making*. (John Wiley, London), pp. 461–84
Pacheco, L. 1972, La inversión extranjera y las corporaciones internacionales en el desarrollo industrial chileno, in Muñoz, O. (ed.) *Proceso a la Industrialización Chilena*. (Universidad Católica, Santiago)
Palma, J. G. 1978, Dependency: A Formal Theory of Underdevelopment or a Methodology for the Analysis of Concrete Situations of Underdevelopment? *World Development* 6, 7/8, pp. 881–924
——— 1979, *Growth and Structure of Chilean Manufacturing Industry from 1830 to 1935*. (Unpublished doctoral dissertation, University of Oxford)
Paquien, J. L. 1969, *La Industria Automotriz en la ALALC*. (INTAL, Buenos Aires)
Pederson, L. R. 1966, *The Mining Industry of the Norte Chico, Chile*. (Northwestern University Studies in Geography, Evanston, Ill.)
Pederson, P. O. 1975, *Urban-regional development in South America: a process of diffusion and integration*. (Mouton, Paris)
Perroux, F. 1950, Economic space: theory and application, *Quarterly Journal of Economics*, 64, Feb.
——— 1971, Note on the concept of growth poles, in Livingstone, I. (ed.) *Economic Policy for development: selected reading*. (Penguin, Harmondsworth), pp. 278–89
Perry, G. 1982, World Markets for Manufacturers and Industrialisation in Developing Countries, in Ffrench-Davis, R. and Tironi, E. (eds) *Latin America and the NIEO*. (Macmillan, London), pp. 126–51
Petras, J. 1969, *Politics and Social Forces in Chilean Development*. (University of California Press, Berkeley)
Pfeiffer, J. B. 1952, Notes on the Heavy Equipment Industry in Chile, 1800–1910, *Hispanic American Historical Review*, 32, 1, pp. 139–44
Philip, G. 1982, *Oil and Politics in Latin America*. (Cambridge University Press, Cambridge)
Pinto, A. 1973, *Chile, un caso de desarrollo frustado*. (Editorial Universitaria, Santiago)
——— and Knakal, J. 1973, The Centre-Periphery System Twenty Years Later, *Social and Economic Studies*, 22, 1, pp. 34–81
———, ——— 1973, *América Latina y el cambio en la economía mundial*. (Instituto de Estudios Peruanos, Lima)
Pirenne, H. 1925, *Medieval cities*. (Princeton)
Platt, D. C. M. 1977, *Busisness Imperialism, 1840–1930*. (Clarendon Press, Oxford)
Portes, A. and Browning, H. (eds) 1976, *Current perspectives in Latin American urban research*. (University of Texas Press, Austin)
——— and Walton, J. 1976, *Urban Latin America: the political conditions from above and below*. (University of Texas Press, Austin)
Prebisch, R. 1962, The economic development of Latin America and its principal problems, *Economic Review of Latin America*, 7, Part 1
Pred, A. 1977, *City Systems in Advanced Economies*. (Hutchinson, London)

Quijano, A. 1974, The marginal pole of the economy and the marginalized labor force, *Economy and Society*, 3, 4, pp. 393–428

Rabinovitz, F. and Trueblood, F. (eds) 1970, *Latin American Urban Research Vol. 1*. (Sage Publications, Beverley Hills)

———, ——— 1973, *Latin American Urban Research Vol. 3*. (Sage Publications, Beverley Hills)

Radice, H. (ed.) 1975, *International Firms and Modern Imperialism*. (Penguin, Harmondsworth)

Richardson, H. W. 1976, The argument for very large cities reconsidered: a comment, *Urban Studies*, 13, pp. 307–10

——— 1980, Polarization reversal in developing countries, *Papers of the Regional Sciences Association*, 45, pp. 67–85

Rippy, J. F. and Pfeiffer, J. 1948, Notes on the Dawn of Manufacturing in Chile, *Hispanic American Historical Review*, 28, 2, pp. 292–303

Roberts, B. 1976, The social history of a provincial town: Huancayo, 1890–1972, in Miller, R., Smith, C. T. and Fisher, J. (eds) *Social and economic change in modern Peru*. (Centre for Latin American Studies, University of Liverpool)

——— 1978, *Cities of Peasants*. (Edward Arnold, London)

Robinson, D. J. 1971, Venezuela and Colombia, in Blakemore, H. and Smith, C. T. (eds) *Latin America Geographical Perspectives*. (Methuen, London), p. 216

Roxborough, I. 1979, *Theories of Underdevelopment*. (Macmillan, London)

Sampson, A. 1981, *The Moneylenders*. (Hodder & Stoughton, London)

Santos, M. 1979, *The Shared Space: the two circuits of the urban economy in underdeveloped countries*. (Methuen, London)

Schmitter, P. C. 1971, Central American Integration: Spill-over, Spill-around or Encapsulation? *Journal of Common Market Studies*, Vol. IX, No. 1, p. 1

Schmitz, H. 1982, *Manufacturing in the Backyard*. (Francis Pinter, London)

Schvarzer, J. 1983, *Cambios en el liderazgo industrial argentino en el período de Martínez de Hoz*. (Centro de Investigaciones Sociales sobre el Estado y la Administración (CISEA), Buenos Aires)

Scobie, J. R. 1971, *Argentina: a city and a nation*. (Oxford University Press, New York)

Scott, M. A. 1976, Who are the self-employed? In Gerry, C. and Bromley, R. (eds) *The Casual Poor in Third World Cities*. (John Wiley, London)

Segre, A. 1982, The Location of the Automobile Industry in Latin America — the cases of Belo Horizonte and Cordoba. (IGU Commission and Industrial Systems Conference, São Paulo)

Sigmund, P. E. 1980, *Multinationals in Latin America: The Politics of Nationalization*. (University of Wisconsin Press, Madison)

Singer, P. 1973, *Economía política da urbanizaçao*. (Edicoes CEBRAP, São Paulo)

Slater, D. 1975, Underdevelopment and spatial inequality; approaches to the problems of regional planning in the Third World, *Progress in Planning*, 4, 2, pp. 97–167

Smith, A. 1970, *The Wealth of Nations*. (Penguin, Harmondsworth)

Smith, C. H. 1980, *Japanese Technology Transfer to Brazil*. (UMI Research Press, Ann Arbor, Michigan)

Smith, D. M. 1971, *Industrial Location*. (John Wiley, New York)

Storper, M. 1984, Who benefits from Industrial Decentralization? Social Power in the Labour Market, Income Distribution and Spatial Policy in Brazil, *Regional Studies*, 18, 2, pp. 143–64

Streeten, P. 1982, A cool look at "outward-looking" strategies for development, *World Economy*, 5, 2, pp. 159–69

Sunkel, O. 1965, Change and Frustration in Chile, in Veliz, C. (ed.) *Obstacles to Change in Latin America*. (Oxford University Press, London)

——— 1969, National development policy and external dependence in Latin America, *Journal of Development Studies*, 6, pp. 23–48

——— 1973, Transnational capitalism and national disintegration in Latin America, *Social and Economic Studies*, 22, pp. 132–76

Taafe, E. J., Morrill, R. L. and Gould, P. R. 1965, Transport Expansion in Underdeveloped Countries: A Comparative Analysis, *Geographical Review*, 53, pp. 503–29

Ternent, J. A. S. 1970, Urban concentration and dispersal: urban policies in Latin America, in Gilbert, A. (ed.) *Development Planning and Spatial Structure*. (John Wiley, New York), pp. 169–85

Thorp, R. and Bertram, G. 1976, Industrialization in an open economy: a case study of Peru, 1890–1940, in Miller, R., Smith, C. T. and Fisher, J. (eds) *Social and economic change in modern Peru*. (Centre for Latin American Studies, University of Liverpool)

Todaro, M. P. 1977, *Economic Development in the Third World*. (Longman, London)

Tornquist, C. 1970, Contact Systems and Regional Development, *Lund Studies in Geography*, Series B35

Townroe, P. M. 1979, Employment Decentralization: Policy Instruments for Large Cities in Less Developed Countries, *Progress in Planning*, 10, pp. 85–154

——— and Keen, D. 1984, Polarization Reversal in the State of São Paulo, Brazil, *Regional Studies*, 18, 1, pp. 45–54

Trebat, T. J. 1983, *Brazil's State-Owned Enterprises*. (Cambridge University Press, Cambridge)

Trebilcock, C. 1981, The Industrialization of the Continental Powers, 1780–1914. (Longman, London)

Tugwell, Franklin. 1975, *The Politics of Oil in Venezuela*. (Stanford University Press, Stanford)

Turner, F. J. 1953, *The frontier in American history*. (New York)

Tussie, D. (ed.) 1983, *Latin America in the World Economy*. (Gower, Aldershot)

Tyler, W. G. 1981, *The Brazilian Industrial Economy*. (Heath, Lexington, Mass.)

United Nations Economic Commission for Latin America (UNECLA) 1966, *The process of industrial development in Latin America*. (New York)

——— 1969, Industrial Development in Latin America, *Economic Bulletin for Latin America*, 14, pp. 3–77

——— 1970, *Development problems in Latin America*. (Austin, Texas)

——— 1979, *Analysis and Prospects of Latin American Industrial Development*. Second Latin American Conference on Industrialization, Cali, Colombia

——— 1979, *International Cooperation for Industrial Development in Latin America*. Second Latin American Conference on Industrialization, Cali, Colombia

UNECLA Joint CEPAL/CTC UNIT, 1977, Las Empresas Transnacionales entre Las Mil Mayores Empresas del Brasil. (Working Paper No. 5, Santiago, Chile)

Uribe Ortega, G. 1967, *La Localización de la Actividad Manufacturera en Chile*. (Universidad de Chile, Santiago)

Vapnarsky, C. A. 1969, On Rank-Size Distribution of Cities: An Ecological Approach, *Economic Development and Cultural Change*, Vol. 17, pp. 584–95

Vernon, R. 1966, International investment and international trade in the product cycle, *Quarterly Journal of Economics*, 80, pp. 190–207

——— 1971, *Sovereignty at Bay: the multinational spread of U.S. enterprises*. (Penguin, Harmondsworth)

——— 1978, *Storm over the multinationals: the real issues*. (Macmillan, London)

——— and Aharoni, Y. (eds). 1981, *State-owned enterprise in the Western economies*. (Croom Helm, London)

Walton, J. 1975, Internal Colonialism: Problems of Definition and Measurement, in Cornelius, W. and Trueblood, F. (eds) *Latin American Urban Research*, pp. 29–52

——— Guadalajara: 1978, Creating the Divided City, in Cornelius, W. and Kemper, R. (eds) *Latin American Urban Research*, 6, pp. 25–50

Watanabe, S. 1970, Entrepreneurship in small enterprises in Japanese manufacturing, *International Labour Review*, 102, 6

Weber, A. 1929, *Theory of the Location of Industries*. (University of Chicago Press, Chicago)

Wells, L. T. (ed.) 1972, *The Product Life Cycle and International Trade*. (Harvard University Press, Boston)

West, P. J. 1977, *The Tyre Multinationals: A Study of Foreign Investment and Technology Transfer in Latin America*. (Ph.D dissertation, University of Sussex)

——— 1978, Venezuela: foreign investment policy, *Bank of London and South America Review*, 12, 3, p. 118

——— 1979, Venezuela: the iron and steel industry, *Bank of London and South America Review*, 13, 3, pp. 138–148

Williamson, J. G. 1965, Regional Inequality and the Process of National Development: A Description of the Patterns, *Economic Development and Cultural Change*, 13, July, pp. 3–45

Wionczek, M. 1981, On the Viability of a Policy for Science and Technology in Mexico, *Latin American Research Review*, XVI, 1, pp. 57–78

Wood, A. 1984, When the bottom goes out of copper, *Geographical Magazine* 56, 1, pp. 16–20

World Bank, 1983, *World Development Report 1983*. (Oxford University Press, New York)

Wythe, G. 1945, *Industry in Latin America*. (Columbia University Press, New York)

INDEX

ABC towns 127–8
Aconcagua 232–4
agglomeration economies 94–6
ALCASA 51
Alessandri 210–11, 219, 220
Allende 62, 214–20
Alliance for Progress 9
Alonso 94, 181
aluminium 41
Andean Group 73, 74, 76–80
  Automobile Programme 76–9
  Decision 24 (foreign investment) 60
  industrial patterns 35–41
  intra-regional trade 77
Angelini group 52
Antofagasta 87–93, 96, 143, 232–3
  role of immigrants 195–6
Arequipa 136
Argentina 82, 84–6
  early industry 22
  industrial structure 35–41
  industrialisation and urbanisation 153, 155–7, 190–1
  manufacturing growth 36
  motor vehicle industry 25–6, 64–6
  multinational companies 64–7
  role of technology 43, 45–6
  state enterprises 50
Arica 89, 95, 152, 211, 219–20, 232
  motor vehicle industry 107–12
  role of immigrants 195–6
  television industry 113
Asunción 136, 139
Atacama 232–3

Barcelona 107, 111, 127
Barquisimeto 127
BAUXIVEN 51–2
Belém 166
Belo Horizonte 98, 107, 111–12, 117–18
Bogota 136, 139
Bolivia 73, 76–80, 86
  industrial structure 37–8
  manufacturing growth 36
  state enterprise 50
Boudeville 106–7

Brazil:
  centralisation 82, 84, 85–6
  computer industry 44–5
  debt 14–16
  early industry 22
  export promotion 8, 28–30, 66
  industrial decentralisation 117–21
  industrial structure 35–42
  industrialisation and urbanisation 153, 155–7, 166–7, 191
  international trade 3, 8
  manufactured exports 8, 19
  manufacturing growth 36
  motor vehicle industry 25–6, 29, 46–7, 64–7, 178
  motor vehicle multinationals 64–7
  National Development Bank 50, 98
  national private firms 54
  Northeast 118–20
    chemical industry 118–20
    metallurgical industry 118–20
  product cycle 31
  regional disparities 179, 181
  role of technology 44–7
  small firms 54
  state firms 50
  steel and metallurgical industries 117, 118–20
Brookfield 159, 163
Buenos Aires 82, 84, 148, 153–6, 186
  historical 136, 139
  primacy 143
Burroughs 44

Caceres 16
Campinas 128
capitalism 181–2
  competitive capitalism 164–6
  monopoly capitalism 164
Caracas 82, 84, 99, 123–7, 160–1
  historical 139
  primacy 144
Cardoso and Faletto 153–4, 156
Casablanca 211
Central American Common Market (CACM) 73–6

General Treaty 75
industrial patterns 35–40
regional allocation of industry 75
Special System for Promotion of Productive Activities 75
centre-periphery model 16–20
Chenery 33
'Chicago boys' 32, 221
Chiguayante 197, 205–6, 231
Chile 73, 76–80, 82, 84–7, 156–7
Allende government 214–20
Nationalisation and intervention 215
requisition 216
supply of money and inflation 217
construction boom 225
debt 16
early industry 194–7
export promotion 19, 32–3
fish-meal industry 225, 232
Frei government 210–4
foreign investment 212–14
regional planning 182–3
relations with United States 212
industrial location 103
industrial stagnation (1950s) 207–10
industrial structure 37–8
manufactured exports 30, 228
motor vehicle industry 25–6, 46–7, 63, 95, 210–14, 223
multinational companies 62–4
Pinochet government 220–34
capital and labour 227
changes in industrial production 223–5
tariff reduction 221–2
protection 23, 197–207
multiple exchange rates 198
Tariff Act 194, 197–8
trade quotas 198
role of immigrant entrepreneurs 195
steel industry 201–2
textile industry 199
trade 62, 228
Chillán 234
China:
comparisons with Latin America 189–90
trade with Chile 228
Chrysler 63, 65–6, 123–5, 127
Citroën 63, 65, 111
Ciudad Bolivar 116
Ciudad Fajardo 124
Ciudad Guayana 114–16, 174
Cobra Computadores 44–5
CODELCO 98, 187, 188
Cohen 186
Colombia 73, 76–80, 86
industrial structure 37–8
manufacturing growth 36
state enterprises 50
tariff levels 32
urbanisation 190–1
Compania de Acero del Pacifico (CAP) 201–2
Concepción region 150, 195, 202, 204, 206–7, 220, 231–3
early industry 197
labour costs and retail prices 87–94
map 205
Congdon 62
Copiapó 140–2
Coquimbo 232–3
Cordoba 107, 111–12
Coronel 205–6
Corporación de Fomento del Desarrollo (CORFO) 156–7, 199–203, 216, 233
Corporación Venezolana de Guayana (CVG) 51–2, 114–16
Corral 197, 201
Costa Rica 73–6, 85–6
debt problems 16
industrial structure 37–8
manufacturing growth 36
Cuba 3
Cuernavaca 127
Cumaná 127
Curicó 232–4
Cuzco 136, 143

Dean 22
debt crisis 13–16, 20
de Castro 63

Economic Commission for Latin America (ECLA) 16–17, 209–10
economic integration 71–81
complementarity 72–3
customs union 72
multinational corporations 74
economies of scale 25–8, 33, 175
Ecuador 73, 76–80, 84, 86
industrial structure 37–8

manufacturing growth 36
EEC trade 7–8, 17
El Salvador 73–6, 86
industrial structure 37–8
manufacturing growth 36
enclave economies 156
ENDESA 201–2
Eurodollar currency market 10
export growth location 175
export promotion 28–34, 61–7
exports:
manufactures 6–7, 8, 17
prices 5
primary products 4–5, 17

Fiat 25, 30, 63–6, 111–12, 116, 125, 127
Firestone 64, 74
Ford 25, 30, 63, 65, 111, 123–5, 127
foreign investment 56–9
Fortaleza 119, 162, 166
Frei 211–14, 219–20
Friedman, Milton 221
Friedmann, John 174–7, 182–91
Furtado 22–3

General Motors 63, 65, 77–9, 123–5, 127–8
Brazil 25, 30
Chile 232
General System of Preferences 7
Gilbert 171, 176, 179
Goodyear 64, 74, 75, 233
Gotuzzo 63, 221
Gradiente 31
growth centres 182–3
growth poles 155
growth pole theory 104–7
Guadalajara 127, 136
Guatemala 73–6, 86
industrial structure 37–8
manufacturing growth 36
Guatemala City, 136, 139
Guayaquil 84
Guri 51, 115–16

Hamilton 22
Hilhorst 184
Hirschman 106–7, 172–4, 175–6
Honduras 73–6, 86
industrial structure 37–8
manufacturing growth 36
'Honduras' syndrome 74
Huachipato 157, 201–2, 205–6, 232
Huancayo 152
Hymer 185–6

Ibañez 220
IBM 44, 99
import substitution 24–8, 49, 85, 209–10
patterns of production 39–41
industrial decentralisation 103–29
industrialisation
and European immigration 153–6
and regional development 171–91
and the state 156–8
and urbanisation 146–67
and urban marginality 163–7
major characteristics 146–8
industrial policies 22–34
INTERALUMINA 51
international capital flows 9–16
International Labour Office (ILO) 162–3
International Monetary Fund (IMF) 10
international trade 3–9
investment and technical change 225–6
Iquique 89, 232

Japan:
multinationals 55
technology 43
trade 7–8, 17, 228
vehicle exports 63
Junta de Adelanto de Arica (JAA) 220

Katz 42–3
Kennelly 84, 117
Kitching 190–1
Kñakal 18
Kuznets 30

labour costs 88–94
labour markets 147–8, 150, 178
La Guaira 124
La Paz 139
La Serena 202
Latin American Free Trade Association (LAFTA) 73
La Urbina 124
Lenin 55, 185
Lewis, W. A. 56
Lima 82, 84, 148, 156, 164
historical 136, 139
primacy 143

Linares 232–3
Lirquén 197, 205–6, 231
localisation economies 94–5
location theory 83–100
Lomé convention 8
Los Andes 108, 110–11, 211
Lota 197, 202, 205–6

Magallanes 195–6
Managua 139
Manufacturas del Cobre (MADECO) 96, 157, 201
Maracáy 124–5
Martinez de Hoz 64, 65
Massachusetts Institute of Technology (MIT) 174, 182
Massey and Meegan 225–6
Matarazzo 154–5
Medellín 136
Mercedes Benz 65, 66, 127
Mexico:
  debt 14
  export promotion 32
  foreign investment 56–8
  industrial decentralisation 117–18, 120–2
    border industrialisation programme 121–2
  industrial structure 35–41
  multinationals 54–8
  private national firms 54
  spatial centralisation 82, 85–6
  state enterprises 50
  steel industry 117
  technology 43–6
  urbanisation 190–1
Mexico City 82, 117, 127, 186
  banking 95–6
  entrepreneurs 98
  historical 136, 137, 139
  primacy 143
migration 97, 177–80
Monclova 121
Monterrey 117–18
Montevideo 84
  historical 136, 139
  primacy 144
motor vehicle industry 25–7, 63–7
  and industrial decentralisation 107–12
  see also country entries
multifibre agreement 7
multinational banks 10
multinational corporations 55–61, 184–8
  and capital flows 9
  and economic integration 74
  and export production 61–7
  and international trade 8
  and the State 59–61
  recent investment in Chile 62–4
Muñoz 211
Myrdal 174–5, 176

national enterprises 52–5
nationalisation 187–8, 215
newly-industrialising countries 19
Nicaragua 3, 73–6, 85–6
  debt 16
  industrial structure 37–8
  manufacturing growth 36
nitrates 142–3
Norte Chico 140–2
Ñuble 232–4

Oaxaca 136
O'Higgins 232–3
oil industry 4, 187
oligopolistic reaction 60
OPEC countries 10–12
Osorno 232–4

Pampas 140, 144
Panama 85–6
  industrial structure 37–8
  manufacturing growth 36
Pâo de Açucar 187–8
Paraguay 84–6
  industrial structure 37–8
  manufacturing growth 36
Paraíba valley 128
Pegaso 125, 127
PEMEX 98, 187–8
Penco 197, 205–6, 223, 231
Peron 157
Perroux 104–6, 155
Peru 56, 73, 76–80, 82, 84–6
  industrial structure 37–8
  manufacturing growth 36
  state enterprises 50
  tariff levels 32
  urbanisation 190–1
Petare 123–4
PETROBRAS 50
PETROVEN 51, 98, 187–8
Peugeot 63, 65
Pinochet government 62–3, 221–34
Pinto 18

polarisation reversal 176–82
pollution 177
Potosí 138
Prebisch 16–18
primacy 176–7
product life cycle 30–1
protectionism:
  developed countries 7
  Latin America 23–4
Puebla 127, 136
Puente Alto 150

Quijano 164–7
Quilpue 207
Quito 136, 139

railways 140–2
Rancagua 108, 110–11, 211
rationalisation 225
Reagan administration 14, 19
Recife 119, 136, 139, 166
regional disparities 171–91
Renault 63, 65
resource growth pole 113–16
retail prices 86–8
Reynolds 51
Rio de Janeiro 84, 98, 136, 138–9, 166
Roberts 152, 158, 166
robotisation 9

Saab-Scania 127
Salvador 119, 136, 139, 166
Sampson, A. 12
San Bernado 150
San José 139
San Salvador 139
Santiago 82, 84, 182–3, 186, 231–3
  early industry 196–7
  historical 136, 139, 143
  immigrants 195–6
  industrialisation and urbanisation 150, 155–8, 164–5
  labour costs and retail prices 87–94
  motor vehicle components 108–11
  spatial centralisation 203–4, 206–7, 219–20
Santo André 127–8
San Vicente 197, 205–6, 232
São Bernardo do Campo 127–8
São Caetano do Sul 127–8
São Paulo 82, 84, 95, 98, 117, 127–8, 171, 178–81, 186
  early industrial growth 22
  historical 136, 140
  industrialisation and urbanisation 153–6, 166
Schmitz 162
secondary city growth 176, 178–80
SIDERBRAS 50, 98
Siderurgíca del Orinoco (SIDOR) 114–16
SIVENSA 98, 116
small firms 54, 160–3
Smith, Adam 22
Sociedad de Fomento Fabril (SOFOFA) 149, 203
Sorocaba 128
state firms 49–52, 187–8
Streeten 33
Superintendencia do Desenvolvimiento do Nordeste (SUDENE) 118–20

Talca 232–4
Talcahuano 197, 205–6, 232
Tarapacá 195–6, 231–3
Taubaté 128
technological change 42–7
Tegucigalpa 139
Tomé 195, 197, 205–6, 223, 231
Townroe 171, 176–80
Toyota 127
trade creation 74
trade diversion 74
Triple Alliance 49, 61
Trujillo 107, 111
tyre industry 60, 64, 74–5

unemployment 165
Unilever 64
United States 181
  Alliance for Progress 9
  budget deficit 14
  early industrialisation 22
  firms 54–8
  trade 7–8
urban primacy 142–4
urbanisation:
  and industrialisation 146–67
  and Spanish empire 136–9
  and the export economy 140–4
  historical perspectives 136–44
urbanisation economies 94–6
urban dual enonomy 158–63
Uruguay 3, 84–6
  industrial structure 36–8

manufacturing growth 36
state enterprises 50

Valdivia 232–4
early industry 197, 201
industrial decline 204, 206
Valencia 115–16, 123–5
Valparaíso 143, 156, 232–3
early industry 197–8, 207, 220
heavy engineering 195
immigrants 195–6
labour costs and retail prices 87–94
Vargas, Getulio 157
Venezuela 73, 76–80, 82, 84–6
Acción Democratica 157
aluminium industry 51
axial belt 115–16, 123–7
export promotion 32
foreign investment 60
import substitution 33
industrial decentralisation 124–7
industrial productivity 33
industrial structure 35–41
motor vehicle industry 25–6, 99–100, 123–5
multinationals 56
national private firms 55
state firms 50–2
steel industry 114–16
Vera Cruz 136
Vial group 53, 64
Videla 64
Viña del Mar 207
Volkswagen 65–6, 77–9, 99, 127–8
Brazil 25, 30, 46
trade 8

Weber, A. 83, 84, 88, 94
Williamson 172–3

Yashica 31